At the Heart of the Coral Triangle

The authors have donated 10% of their royalties to Conservation International's Coral Triangle Initiative (CTI). Learn about the CTI's work to combat threats from climate change and unsustainable fishing to support this extraordinary environment at https://www.conservation.org/projects/coral-triangle-initiative

At the Heart of the Coral Triangle

Celebrating Biodiversity

CRC Press
Taylor & Francis Group
Boca Raton London New York

CRC Press is an imprint of the
Taylor & Francis Group, an **informa** business

First edition published 2021
by CRC Press
6000 Broken Sound Parkway NW, Suite 300, Boca Raton, FL 33487-2742

and by CRC Press
2 Park Square, Milton Park, Abingdon, Oxon, OX14 4RN

Photographs and Text © Alan J Powderham

CRC Press is an imprint of Taylor & Francis Group, LLC

Reasonable efforts have been made to publish reliable data and information, but the author and publisher cannot assume responsibility for the validity of all materials or the consequences of their use. The author and publishers have attempted to trace the copyright holders of all material reproduced in this publication and apologize to copyright holders if permission to publish in this form has not been obtained. If any copyright material has not been acknowledged please write and let us know so we may rectify in any future reprint.

Except as permitted under U.S. Copyright Law, no part of this book may be reprinted, reproduced, transmitted, or utilized in any form by any electronic, mechanical, or other means, now known or hereafter invented, including photocopying, microfilming, and recording, or in any information storage or retrieval system, without written permission from the author.

For permission to photocopy or use material electronically from this work, access www.copyright.com or contact the Copyright Clearance Center, Inc. (CCC), 222 Rosewood Drive, Danvers, MA 01923, 978-750-8400. For works that are not available on CCC please contact mpkbookspermissions@tandf.co.uk

Trademark notice: Product or corporate names may be trademarks or registered trademarks, and are used only for identification and explanation without intent to infringe.

ISBN: 978-0-367-42816-7 (hbk)
ISBN: 978-0-367-85531-4 (ebk)

Typeset in Times NR MT Pro
by KnowledgeWorks Global Ltd.
Graphic design by Alan J Powderham

ACKNOWLEDGEMENTS

The authors would like to thank the numerous friends and colleagues for their specialist advice on the marine biology and in particular: Gerry Allen, Oliver Crimmen, Sammy DeGrave, Mark Erdmann, Daphne Fautin, Bruno Hopff, Harry ten Hove, Rita Mehta, Gustav Paulay, James Reimer, Noa Shenkar, Rob Sinke, Alex Vagelli, and Nicole de Voogd. Special thanks are also due to Liz Powderham and Bastian Reijnen for their support, patience and insight, and to Laura Powderham for her advice and original inspiration for the graphic design.

CONTENTS

Foreword viii

Introduction x

Chapter 1
Seascapes
Passing perfection? 2
The deep diversity of the shallow reef 8
Uncommon gardens 10
Anemone city 14
Life's a beach 16
Safe haven 18
The sounds of silence 22
Finding a sweet spot 24
School time 26

Chapter 2
Fish portraits
Intelligent life 30
The sunfish's tale 32
Captivating mimicry 34
Colourful confusion 36
Pigment patterns 38
Trigger happy 42
Boxing clever 44
Midas touch 46
Horse sense 48
Pygmy seahorses 50
Manta rays 52
The wonders of whale sharks 54

Chapter 3
Invertebrates
Lords of the rings 60
Fatal attraction 62
Stars and stripes 64
Marbled exotica 66
Sitting tenants 68
Eccentric shrimps 70
Beauty and the beast 72
Antennae range 74
Changing the guard 76
Aliens 78
Poetry in mantles 80
The lost chord 82

Chapter 4
Predation
The first link 86
Flying carpets 90
Bubble netted jacks 92
The art of ambush 96
Goby dessert 98
Sneak attack 100
Twice bitten 102
Fishing fans 104
Legless in Ambon 106
Stunning weapons 108
Night watch 110
Strange intelligence 112

Chapter 5
Reproduction
Naked ambition 116
A brittle embrace 120
Anemone invasion 122
Endangered yet invasive 124
Plagues and pufferfish 126
Pipe dreams 128
Replicating rainbows 130
Flasher frenetics 132

Chapter 6
Behaviour
The shape of water 136
Manta opportunists 138
Juvenile behaviour 140
Hide and seek 144
Dancing classes 146
Shape shifters 148
Mercury rising 150
Coconut octopus 152
See-through security 154
Vanishing act 156
Dressed to the nines 158
Pirates and bonsai anemones 160

Chapter 7
Symbiosis
Constructive synergy 164
Ménage à trois 166
Gardening in a ring of fire 170
Candy crabs 172
Symbiotic embedments 174
Clinging together 176
Taken to the cleaners? 178
Cleaning interrupted 182
An odd couple 184
A double mystery 186

Chapter 8
Reptiles
A breath of fresh air 190
Resident evil 192
Turning turtle 194
Becoming vegan 196

Chapter 9
Conservation
Climate change 202
Destructive fishing methods 204
Unsafe harbour 206
Exotic invaders 208
The small majority and Darwin's Paradox 210
Beneath the volcano 212

Appendix
Abbreviations 215
Equipment 215
Maps 216–217
Glossary 218–221
Photo guide 222–234
Further reading 236–238
Index 240–241

The coral reefs of the tropics are a marine ecosystem for which there are many superlatives. They contain more species than any other, not just those organisms large enough for a diver to see, but with myriad small creatures in crevices and holes where they make homes, hunt for prey, avoid becoming prey, and hunt for mates too. Reefs are highly productive, producing vast amounts of surplus protein which has benefited people for millennia. They produce the rocky foundations that rise up to the surface that can be seen from space. The corals, which are at the core of the coral reef, are the only organisms that make land—in some cases, entire island states—again, to the benefit of people.

Sadly, however, these are bad times for coral reefs. Most of the key threats derive from ever-growing demands and activities of world's increasing population. Coral reefs are extensively polluted from industry, including sewage and fertiliser run-off from farms. Widespread deforestation on land causes copious amounts of sediment inundating river systems, which on reaching the sea overwhelms coral reefs. Reef building corals are being bleached and killed on a catastrophic scale by climate change driven by rising greenhouse gas emissions into the atmosphere. Coral reef ecosystems are being increasingly exploited by taking too many fish and other creatures too. This critically damages the ecosystems even in places where there is no pollution. At this moment, about half of all the reefs in the world have been damaged or killed. It is not a good time for coral reefs.

We are causing these problems, but the solutions to protect and recover coral reefs also lie within our hands. There is always hope and it is certainly never too late to stop trying. Over half a billion people depend on reefs for their protein and for shoreline protection because reefs form natural breakwaters that stop our coastal infrastructure from being washed away. Livelihoods, even lives, depend on healthy coral reefs and in the more prosperous countries, trillions of dollars of infrastructure remains intact because of them. A common political response from our leaders who lack sufficient foresight is to say we cannot afford to do anything about it. But, in fact, we cannot afford not to.

One of the most important aspects of coral reefs is their beauty. An appreciation of beauty is a key attribute that can elevate people above other species, above and beyond the basic requirements of feeding and reproducing sufficiently to sustain our populations. People see beauty in many ways—in a child, in a wise and loving grandmother, in a partner, in paintings or music. From the perspective of a scientist, this beauty is also evident within the mosaic of connected forms and living patterns that make up an ecosystem, particularly one as elaborate as a coral reef. Most ecosystems have their own unique attributes, but coral reefs have more species and so more interdependent components than any other, and it is a joy to try and unravel and understand their nature. As a marine system, the species of a reef, especially the heavy skeletons of corals, are buoyed up by water so they become largely immune to the effects of gravity, allowing corals to develop intricate three-dimensional structures that are so important to the reefs' rich life. Coral reefs are an ancient ecosystem—far older than we humans—and their components often look bizarre to us, with the strangest of shapes, behaviours and psychedelic colours and patterns.

Photography can provide a compelling way to convey the extraordinary nature of coral reefs, especially to the vastly greater number of people who will never be fortunate enough to visit one. Photography is also an effective means to convey their importance—by capturing their appearance, showing how their component species interact with each other and, indeed, how they interact with us. Scientists have been trying to promulgate the value of reefs for several decades along with warnings of how and why losing them will harm us all, yet the devastation continues. We need to do more to conserve them. One might think that this would not be too difficult given that self-interest is involved, but a marine ecosystem far away in the tropics is out of sight and so out of mind for most people. This publication will help stimulate awareness of coral reefs for many more.

he focus of this book is on Southeast Asia's reefs in the 'Coral Triangle'. These are at the centre of global biodiversity, with more div
sh and invertebrate life than anywhere else in the world. More than that, it is the region of origin of many species that have subsequen
est, north and south from there. Being the location of many archipelagos—mainly of the Philippines, Malaysia and with eastern Indo
egion has an enormous coastline which is one reason for its diversity. It is also a highly populated part of the world, where reefs that use
le can no longer do so. With this diminishing ability of the inhabitants to secure their food or a living, this can exacerbate many of the

We need to significantly raise awareness because reefs today face an existential threat. But, we cannot do so effectively until a wider p
nformed and joins in too. Because most people will not see, let alone study and understand coral reefs, we need books whose photogr
heir beauty, nature and extraordinary species. Reefs are the canaries in the coal mine in the sense that they are an ecosystem that is
he global extinction of species and systems that began ever since we entered this uncharted Anthropocene epoch. The stunning pho
howing images of some of the world's richest and most intact coral reefs will help to raise awareness, and that is going to be key.

Charles Sheppard OBE
Professor Emeritus

ving journeys and, in a way, this book is the culmination of one that started over 40 years ago in the Caribbean. Although the
rably less marine biodiversity than the Indo-Pacific region, in those days its coral reefs were thriving with little hint of climate change
The book itself touches a more timeless journey, commencing with visions of pristine coral reefs and concludes with a primeval driver
olcano. In this, it reaches beyond a contemporary perspective into the dimension of deep time and how the very existence of the Coral
ver 20 million years ago.

rney is set out in nine chapters, each with a leading theme in which each subject may be read as self-contained, but which also interre-
nd across the chapters.

y journey took me on to the Red Sea, followed by the Maldives and eventually to the Coral Triangle in 2004. Each region brought its own
nce of the Coral Triangle was immediately breath-taking. Once discovered, I have never wished to dive anywhere else. With such a spectacular
across an area of 6 million square kilometres, there was no risk of a lack of variety. Yet, retracing those initial steps in the Caribbean prompt-
asy it is to forget how alien the marine environment is especially to non-divers. With adequate training, one may enter this strange world
mparative ease. However, it is quite another challenge to really appreciate its wonders. They were far from revealed on my first dive back in
was all about just being immersed in the bizarre act of breathing underwater. The experience was physical and inward rather than about the
exhilarating, but was tempered by the concern of whether I would be able to manage the cumbersome technology of scuba. The heavy tank
ht of the other equipment was impressive. I wondered whether the weight belt with several more kilogrammes of lead was really necessary. I
hard it was to get below the surface when I had half expected to sink like a stone. Looking back thousands of dives later, I realise how fortu-
v this extended journey. In those early days, we did not have the integrated sophistication of modern equipment, so there was a steep learning
buoyancy. That achieved, I took up underwater photography later that year. Another key difference was the medium of film. This meant that
f 36 shots per dive and less if a partly finished roll was carried over to the next dive. This imposed the need to make each shot really count.
n unwelcome constraint, particularly when interesting encounters occurred after the last shot was taken. This happened more often than one
eemed to instinctively know when there was no film left, so I soon learnt to keep a couple of shots right to the end of each dive. Curiously, the
ters then noticeably reduced. Fish are intelligent, but I do not think that they count the shots. Rather, I suspect they rely on body language,
n one has genuinely run out of film. In retrospect, I now appreciate how important that limitation imposed by film proved to be. It taught
axed and honed my skills in observation. This brought real benefits not only to my photography but also to the overall experience of diving.
heras, it is quite easy to take hundreds of shots during just one dive. There is a risk of missing the real experience by spending most of a dive
r checking the results on the back-screen display.

ise that you can actually get quite cold when diving in the tropics. This particularly applies where upwelling currents bring cold water
is 800 times denser than air and its physical properties present plenty of challenges, especially to the underwater photographer. The
water is 20 times that of air making it a very effective heat sink. This, combined with evaporative cooling while breathing compressed
causes continual heat loss during a dive. I did not appreciate how significant this was until I started diving with a rebreather. Because
are literally rebreathing your own air at ambient pressure. Aided by the chemical reaction of removing the exhaled carbon dioxide,

non-intrusive but it also keeps you warm. The multiple benefits of using a rebreather are highlighted in many of the subjects in thi important to avoid getting too cold while diving. It can impair concentration and is hardly conducive to relaxed observation of ma which often requires keeping still for extended periods. I have waited to capture aspects of behaviour until, with my body temperatu arms began to shake. It has proved necessary to move away for an energetic swim to warm up with the hope that the subject woul return. The surface temperature of the sea in the Coral Triangle is typically around 28–29°C. This is perfectly pleasant for diving eve However, with a drop of just a couple of degrees, it starts to feel noticeably cold while entering water at 22°C delivers quite a shock. immediately starts to seep into your wetsuit making you shiver until your body warms this influx. You still feel cold though. Whe cries of the dive guides as they enter to make a pre-dive survey, you know that it is worth adding an extra layer of wetsuit.

Finally, I would like to express my appreciation to Dr. Sancia van der Meij, who has made such a major contribution through he review. Given the range of subjects addressed, this has been a substantial challenge and one she has met with energy and percep underestimated for it is far easier to ask questions than to answer them. It has been a real pleasure working together, exploring a mu insights to the incredible marine life of the Coral Triangle. She has also brought engaging sense of humour, especially when it was n of my more speculative efforts to explain unusual behaviour of marine life.

Alan Powderham

Passing perfection?

The 'Coral Triangle', encompassing the zones of maximum marine biodiversity on the planet, was formally recognized in 2009 at the World Ocean's Conference at Manado in Sulawesi, Indonesia. It embraces an area of 6 million square kilometres extending east-west from Malaysia to the Solomon Islands and from the Philippines in the north to Indonesia and Timor-Leste in the south (see the maps in Appendix, pages 216 and 217). The true depth of its diversity is still being discovered and appreciated and it has been evocatively described as the Eden of oceans and a crucible of evolution. Its geological history was initiated some 23 million years ago as the tectonic plates of the continent of Australia collided with those of Asia creating tens of thousands of islands thrust up from the barren sea floor by the resulting volcanos. The key role of volcanos in the formation of coral reefs is further discussed in the section on Komba in Chapter 9. The Coral Triangle now forms the junction between the Indian and Pacific Oceans, thus sharing and dispersing the biodiversity of both of these vast regions. Indonesia is by far the largest country in the Coral Triangle, with its eastern third, extending some 1500 kilometres from Bali to West Papua, sitting right at the centre. This book is focussed on that central zone and highlights the rich and varied habitats of its marine life.

Our journey within the Coral Triangle commences with an exemplar of coral reefs—the Kusu Ridge. This extensive formation lies in the passage between the islands of Halmahera and Bacan. Its reefs present a vibrant example of biodiversity. Owing to its geographical location, the area is subject to strong currents. While these provide a rich supply of nutrients, the area, like much of the Coral Triangle, is also subject to heavy storms. So, it is a testament to the health of these coral reefs that, despite these potentially adverse natural forces, they extend to within a metre of the surface. As shown opposite and to the left, a variety of seemingly quite fragile structures of *Montipora* coral form high vertical stands that display no evidence of damage. In fact, the whole extent of the reef down to 30 metres appears in excellent condition. The range and variety of these hard coral gardens is of quite breathtaking beauty, as further illustrated by the two panoramas of its near surface reefs on the following pages. Viewing the coral gardens of Kusu Ridge conveys an overwhelming sense of natural perfection combined with an inherent vulnerability. The resonance of this duality is echoed in the 'The deep diversity of the shallow reef' and 'Uncommon gardens' in the following sections. While they illustrate how spectacular, richly diverse and healthy coral reefs can be, such examples are unfortunately becoming increasingly rare throughout the tropics.

Location: Pulau Kusu, South Halmahera, Maluku

The deep diversity of the shallow reef

The special glory of the Coral Triangle is expressed through the spectacular beauty of its shallow reefs. Even with the knowledge of the scientific background described below, the vibrant richness and variety of form and colour are both unexpected and intoxicating. This visual feast is echoed in the series of photographs in these and the following pages. Apart from the wondrous range of corals, these reefs provide sanctuary and nurseries for an immense range of fish and invertebrates. Though far less famous than other locations renowned for tropical coral reefs, such as the Great Barrier Reef (GBR), the region embraced by the Coral Triangle contains some 30% of the world's tropical reefs. It is also the richest centre of marine biodiversity on the planet. It includes over 75% of the known species of corals and nearly 40% of the world's species of coral reef fish. Scientific studies have shown that much of this biodiversity is concentrated in the top 25 metres. Surveys of fish on the deep and shallow reefs in the Coral Triangle have revealed surprising divergence with those in the Pacific. While around 500 species of fish have been recorded on the deep reefs down to around 100 metres in both locations, the shallow reefs of the Coral Triangle, with around 2500 species, harbour more than five times the number encountered in the shallow habitats of the Pacific. Establishing the reasons for this incredible diversity in marine life is a continuing area of study by scientists. Now, with the ever-growing threat from climate change, achieving this understanding has become critically important. It is vital to know how such a uniquely rich biodiversity has evolved and, crucially, why it has been significantly more sustained in the Coral Triangle compared to other major coral reef centres which are now proving less resilient and particularly vulnerable.

During 2014 to 2017, the adverse effects of the El Niño climate cycle were exacerbated by the increased temperatures from global warming. This resulted in the worst coral bleaching events so far recorded in the world's tropical reefs—dramatically devastating vast areas of the GBR and the reefs of the Chagos Archipelago. And yet, surveys found the reefs at the centre of the Coral Triangle in Sulawesi to be surprisingly healthy. The photos featured here were taken in locations some 1000 kilometres further east in Indonesia during this period and also show rich and healthy reefs. The Sulawesi research programme is part of the '50 Reefs Initiative'. This is one of the key projects focussed on answering the questions of coral reef resilience and identifying which species are most likely to survive in the face of climate change.

Location: South Raja Ampat, West Papua

Uncommon gardens

Tropical coral reefs have been a source of fascination and awe for centuries. The nature of their evolution has preoccupied many naturalists—most famously Charles Darwin and, in the Coral Triangle, Alfred Russell Wallace. Darwin's voyage on the Beagle during the 1830s proved pivotal in developing the science of coral reefs and their geologically driven formation. Many mysteries still remain to be fully resolved, including how such unrivalled rich and concentrated biodiversity could be sustained in the apparently nutrient poor waters of coral seas. This became known as Darwin's Paradox (see discussion on this and the geological factors in coral reef formation in Chapter 9).

Coral reefs are often referred to as coral gardens. Certainly, with their evocative shapes and vibrant colours, they can appear quite garden-like. Appearances, of course, can be deceptive and those of coral reefs are charmingly so. Real gardens are an artifice of our making and, while undoubtedly attractive, can hardly compare with the beauty and seemingly endless variety of coral reefs. These are natural creations and, despite our perceptions, are not full of exotic plants swaying from outcrops of exquisitely patterned rocks. Instead, each structure is built by colonies of tiny animals—the coral polyps. Each one, often less than a centimetre in size, is set in a limestone base, which forms the hard structure of the coral reef. The polyps have a tentacled arrangement armed with stinging cells, called nematocysts, which are used in the capture of their planktonic prey. Yet, while not gardens of the terrestrial kind, coral reefs do, in fact, contain an abundance of plant life. It is everywhere but concealed in plain sight. Absorbed within the transparent tissue of the corals, it also provides their colour. In a remarkable symbiosis, microscopic single-celled algae known as zooxanthellae live and photosynthesise within the coral tissue providing its host with vital nutrients. There can be millions of such cells in just one square centimetre. They are also critical in limestone deposition enabling corals to create structures as vast as the Great Barrier Reef. When corals are seriously stressed, for example, during sustained rises in sea temperature, they eject their algal cells revealing the white limestone structure beneath; hence, the term 'coral bleaching'. While the tissue in most species of sea fans and other soft corals is also relatively transparent, their rich colours do not derive from their symbiotic algae, but from that of their underlying structure. Further, 'garden' examples are shown overleaf, including a combination of hard and soft corals, which creates a medusa-like appearance.

Location: Misool, Raja Ampat, West Papua

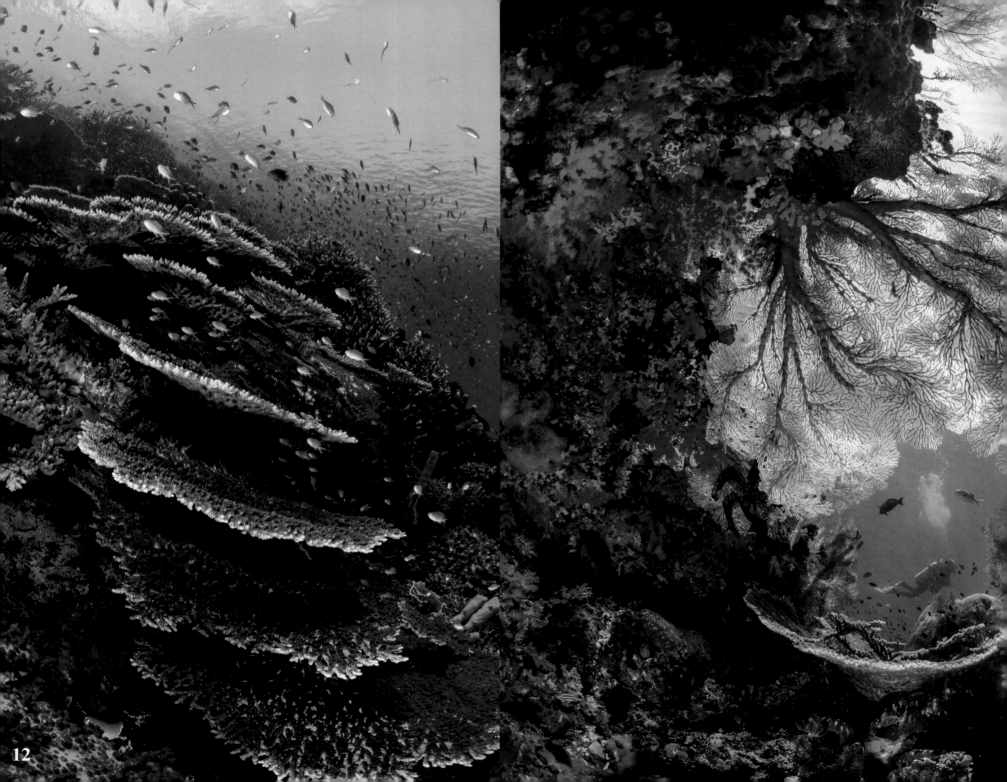

Anemone city

This is a very distinctive and, at least in its coverage, probably unique reef panorama as it features an almost complete dominance of sea anemones. Here, across wide expanses of the very extensive reef, only a few isolated examples of coral are evident. Instead, lush fields of anemones extend, with hardly a break, over many square kilometres, ranging in depth from 10 to 20 metres. The dominant species is the bubble-tentacle sea anemone, but many other species are also present. The surface water temperatures at this location in Alor are usually around 22°C. This is some six to eight degrees less than the normal average on coral reefs in the region. The location is subject to very strong currents, which, at peak flow, can reach over eight knots. This, combined with the underwater topography, creates substantial upwelling of the colder water from below the thermocline, which is typically found at a depth of 40 metres. Such conditions make for advanced level diving and the site can only be tackled comfortably around slack tide. Even then the currents tend to be quite significant, essentially making it a drift dive. This enables a wide area to be easily traversed in one dive, revealing the impressive topography of the anemone fields.

While strong currents are a common feature in the coral reefs in the Coral Triangle, the combination with such low temperatures is more unusual, though far from unique. These conditions generated by strong upwelling currents are also encountered, for example, in Nusa Penida at the south east corner of Bali. However, none of these other locations has anything like a similar abundance of anemones. Why they should thrive so well in Alor Strait and so comprehensively dominate the reefs is unclear. They certainly appear to be very healthy even though the majority of these anemones seem to be without their usual complement of associate anemonefish. They are present, but not in sufficient numbers to match the abundance of anemones. Such an extensive occurrence of healthy anemones flourishing without the apparent support of anemonefish merits scientific investigation. One of the other common associations also absent here is that of anthias with corals. The currents bring a rich supply of planktonic food; so, to exploit this bounty, the anthias here are obliged to adopt whatever alternative sources of shelter are available. These include sporadic sponges as shown in the top left photo and, in the lower one; some anthias have even sought refuge within living clam. It can be seen that both of these sanctuaries sit surrounded by a field of anemones.

Location: Sumba Strait, Flores

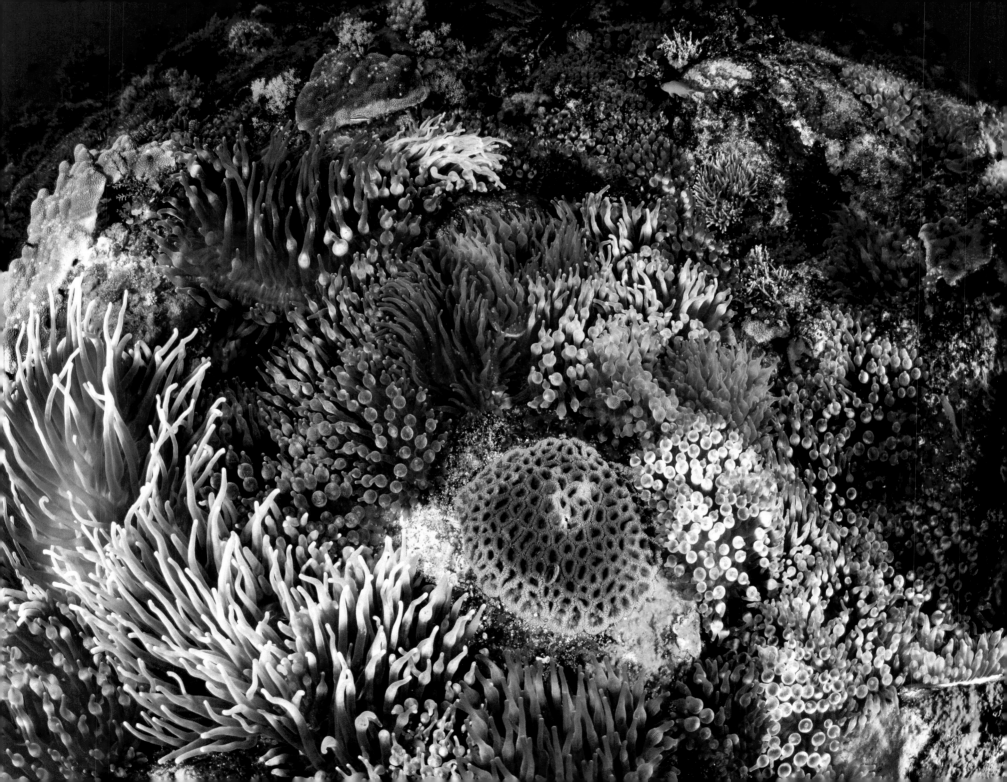

Life's a beach

With few exceptions among their 90 or so species, parrotfish are herbivores. Yet, it could be said that the life of a parrotfish is dedicated to the production of beaches—those beautiful white sandy beaches that we romantically associate with tropical coral islands. So, although essentially vegetarians, parrotfish have evolved a dentition and digestive system perfectly suited to the generation of vast amounts of sand.

The Bumphead parrotfish is the largest species and may attain a length of 1.2 metres and weigh over 50 kilograms. An adult is shown opposite passing copious amounts of sand. Parrotfish graze on the algae that cover coral and other calcareous substrates. They achieve this either by surface abrasion through scraping or, in the larger species, by literally breaking off chunks of substrate. Consequently, parrotfish inevitably consume vast quantities of indigestible calcareous material. Their name is derived from the distinctive beak-like fusion of the front teeth as shown in the close-up of a sleeping Ember parrotfish on the left. Their pharyngeal teeth are also adapted to grind the ingested material down into fine particles and, instead of a true stomach, they have an elongated gut, which is ideally adapted to digesting plant matter and excreting sand.

A healthy reef supports a large population of parrotfish and it has been estimated that just one fish can produce up to a ton of sand every six months. In fact, their combined efforts are believed to generate over 85% of all the sand on a coral reef. Thus, in an ironic paradox, those idyllic beaches are essentially created from parrotfish excrement. More importantly, parrotfish are the leading producers of island-building sediment. Through this, they deliver a vital ecological contribution—one with additional importance given the increasing trend in sea level rise. The photo below shows a pair of Peacock or Flowery flounders enjoying the haven of an underwater beach.

Location: Pulau Koon, North Banda Sea

Safe haven

Compared with the sad state of Ambon harbour as described in Chapter 9, this small bay, with its timber jetty in Raja Ampat, appears relatively pristine. There was no obvious pollution and the marine life seemed healthy. Such structures are a magnet for fish both as a sanctuary and here, for the visiting shoal of Pinnate batfish shown opposite, as a cleaning station. Two small yellow- and black-striped fish with the tinge of blue can be seen near the centre. They are cleaner wrasse and should be distinguished from the members of the ubiquitous shoal of similar-sized fish that have also invaded the space beneath the jetty. These are juveniles and their black- and white-striped livery reflects their common name of Convict blennies. By maturity, their appearance has changed to become even more convict-like though the adults are very rarely seen. This is a little surprising since large aggregations of the juveniles are often encountered on the outer reef slopes. To find them so close to shore and in shallow water as shown here is less usual. It further underlines the attractiveness of these refuges. However, from the batfishes' perspective, the sudden arrival of the blennies may have interrupted the cleaning service. The batfish, though, clearly remain optimistic as individuals can be seen changing colour shades from dark to light—a typical signal that they wish to be cleaned. Further details of cleaning and cleaning stations are given in Chapter 7 on symbiosis.

The species in the two shoals featured here have contrasting ecologies. While, as noted, large shoals of juvenile Convict blennies are common, sightings of juvenile Pinnate batfish are fairly sparse. The juveniles are usually solitary though, as indicated in the left hand photo, they are willing to share territory with small species such as cardinalfish. Their striking colour scheme combines with a mesmerising undulatory swimming action. It is a form of Batesian mimicry in which these juveniles convey the appearance and motion of a foul-tasting flatworm as shown in the lower photo. The fish will further enhance the deception by lying flat and swimming sideways. As it matures, it loses its exotic appearance. The beginning of that metamorphosis is evident in the individual pictured here as its colour is already changing to the silvery hue of the adult fish. Other species of batfish exploit radically different forms of mimicry as described in 'Juvenile behaviour' in Chapter 6. Adult batfish are commonly seen in quite large numbers gathering in open water above the reefs. Further examples of shoaling juvenile Convict blennies and adult batfish are shown in the two following pages.

Location: Sauwandarek Jetty, South Pulau Mansuar, Raja Ampat

The sounds of silence

'The silent world' was famously coined at the advent of scuba diving in the 1940s by Jacques Cousteau to convey the novel and incredible experience that it offered. In reality, with standard diving equipment, any sense of true silence is only momentary between breaths. And even so, the underwater world is hardly devoid of sound. Fish make many vocalisations; parrotfish are continuously scraping away at the coral and even a turning school of fish emits a soft swishing sound like a flock of birds. Apart from volcanos, the loudest natural sound I have heard underwater is that from the tail of a large grouper when it strikes and is literally like the crack of a whip. However, these natural sources are far exceeded by those of breathing from diver. To be really immersed in the quasi-silence of the reef, one needs a rebreather. As noted in the Introduction, this equipment brings many benefits. Having a closed breathing circuit, these eliminate the noisy and disruptive exhalations of bubbles which are inevitable with normal scuba. They enable intimate and extended observations of marine life and bring a heightened awareness of becoming one with the underwater environment.

No other experience quite compares to floating in neutral buoyancy, immersed in the living tableaux of rich reefscapes. In a glittering kaleidoscope of movement, such schools of Golden sweepers can become almost mesmeric. Although these scenes were impressively crowded with fish, they conveyed a seductive sense of peace and serenity. The sweepers featured here are just a few centimetres long and belong to a family of fish, which is generally confined to the Indo-Pacific region. There are around 25 species and, some being quite transparent are also known as glassfish. They are generally most active at night and have the large eyes typical in nocturnal feeders. During the day, they can be observed in closely packed aggregations as they shelter in caves or beneath shady overhangs. They may share such sanctuaries with other fish, even much larger species, as shown in 'Finding a sweet spot'. This contrasts with the diurnal zooplankton feeders which are seen on the right, high up in the water column above the fan of *Acropora* coral beneath which the dense school of sweepers crowd for protection. During periods of low visibility and abundance of zooplankton, the sweepers may be tempted out to feed during the day as shown on the left as they rise from their refuge in an upward spiral into open water.

Location: Misool and Triton Bay, West Papua

Finding a sweet spot

These densely packed aggregations of Ribboned sweetlips were located at a depth of 30 metres. They are nocturnal feeders and are generally encountered in small groups during the day where they are dispersed around the reef. To find so many gathered together in excellent visibility and with such attractive geometric formations was a special pleasure. Fish are very adept at energy conservation, as in these examples, by exploiting sheltered places to rest during their diurnal repose. The fish were calmly crowded into deep crevices in the reef well out of the main tidal flows. This enhanced the photographic opportunities since it was possible to gently and slowly approach very close to the fish without alarming them. When the current does pick up, the fish will then align together and head into it. They, thus, synchronise their formations to convert from the loose arrangement of a shoal into the harmonised coordination of schooling, as explained in 'School time' later in this chapter. Even though, in this case, there was very little current, the efficient fluid dynamics of schooling enabled the tight formations to be maintained. The closely aligned grouping of schooling fish is evident in the two right hand photos, while that on the left, in a virtually current-free shelter, shows two distinct but very inter-mingled shoals—one of sweetlips and the other of Golden sweepers. Although crowded together, the fish are rather randomly arranged with many pointing in different directions. The two species, while contrasting in size, probably make quite compatible groupings since both species are nocturnal feeders but do not compete for the same food. The sweetlips feed on a variety of benthic invertebrates, while the sweepers consume zooplankton as noted previously in 'The sounds of silence'.

The conditions provided great photographic potential. However, to capture these vivid images with a high depth of field demanded a fisheye lens and involved placing the dome port of my underwater camera housing literally within an inch or so of the leading fish. These groups will readily disperse before tolerating such close proximity. Just one exhalation of bubbles can completely disrupt such a scene and it will typically take a good 30 minutes before the fish return to fully regroup and compose. The use of a rebreather enabled the sublime experience of these encounters and the success of the photography.

Location: North Raja Ampat, West Papua

School time

One of the common misnomers about diving and fish is expressed by the adage: 'No current, no fish'. It is a frequent refrain cited in pre-dive briefings. This derives from the assumption that, without current to concentrate and orientate them, fish will be randomly and widely dispersed all over the reef. Certainly, fish readily gather in currents to feed when they are conveyors of food such as plankton or smaller fish. And, indeed, congregations of feeding fish can be quite spectacular. But, like other animals, fish naturally avoid unnecessary expenditure of energy and will not doggedly persist in swimming against currents when there is little or no benefit. All animals need to conserve energy and free-swimming fish will seek respite from currents while not actively feeding. This is immediately evident with nocturnal feeders at rest during the day as related in 'Finding a sweet spot'. The absence of current can also stimulate circular formations as displayed by the barracuda in 'The shape of water' in Chapter 6. Such formations are present in both of the photos shown here, and they also feature dense intermingling aggregations of different species of fish. Yet, these examples were encountered between tides in conditions of zero current. Contrary to the accepted view, it is during such periods of slack water that the greatest concentrations of fish may occur. When diving at the same location on other occasions, but in currents ranging from mild to strong, considerably less concentrations of fish have been evident—even with the benefit of using a rebreather.

But, what makes fish form schools as opposed to shoals and what is the difference between these behaviours? Shoaling applies when a particular fish species simply gather together without any specific overall orientation. The photos show examples of this. Schooling, on the other hand, involves closely coordinated movements between neighbours to achieve synchronised swimming of the whole group. This tactic is also seen in many herding animals and some bird species such as displayed in the well-known 'murmurations' of starlings, and serves to confuse and potentially deter predators. Recent research has indicated that schooling in fish is evolved rather than learned behaviour and one that brings other benefits such as energy conservation. And, while currents act as catalysts to convert shoals into protective schools, fish also school in still water since, most importantly, it concentrates breeding fish together, and thus, enhances the chances of successful reproduction when spawning.

Location: Komodo National Marine Park, Lesser Sunda Islands

CHAPTER 2

FISH PORTRAITS

Intelligent life

Historically, fish have often been regarded as evolutionarily primitive with low intelligence and 3 second memories yet, countless observations of marine life have consistently shown that fish are notably intelligent and naturally curious. They have successfully evolved over 400 million years and, with more than 34,000 species, have achieved a diversity exceeding that of all other vertebrates combined. The photos here show Blacktip reef sharks, barracuda, and Dart trevallies approaching in close proximity to investigate my presence. It has been a constant source of delight and fascination to encounter such overt interest, an experience greatly enhanced if I remained patiently still in one location. Quite soon I would become more the observed than the observer. The inherent curiosity of fish is even more evident when diving with a rebreather. While my fellow divers using standard scuba gear merely received fleeting visitations, at best, they have reported that I often disappear completely enveloped within concentrated shoals of fish. Without the deterrent of noisy exhalations, this behaviour became so predictable that I could select an attractive reef panorama and simply wait for the arrival of the fish to complete the picture. An example of this is shown in 'First link' in Chapter 4, where massed shoals of fusiliers can be seen crossing above a red barrel sponge.

As noted later in this chapter, the majestic manta rays are widely noted for their interest in divers, but similar curiosity is shown by the smallest to the largest of fish. An amusing example occurred while quietly trying to photograph blennies. A pair of these shy fish is shown in 'The small majority and Darwin's Paradox' in Chapter 9. The Red coral grouper, shown in 'The deep diversity of the shallow reef' in Chapter 1, tend to be even more evasive and difficult to approach for a good portrait. On the occasion with the blennies, I had the sudden sense of also being closely watched. Turning to look back I was greeted by the sight of no less than five of these groupers hovering so closely behind me that they could peer over my shoulder—apparently quite absorbed with my activities. To my amazement, they did not immediately disperse upon my discovery, presumably having concluded that I was not a threat. Fish are indeed surprising and curious.

Location: North Raja Ampat, West Papua

The sunfish's tale

The relationship between form and function has been much debated since Aristotle and deeply considered by Darwin. The Oceanic sunfish, *Mola mola*, presents an interesting enigma. Although rarely seen on coral reefs, these gigantic fish make an unforgettably dramatic encounter at close quarters. They are the heaviest of all bony fish. The largest recorded was some 3 metres long and weighed nearly 2.5 tonnes. They produce more eggs than any other vertebrate—exceeding 300 million at a time. This is all achieved on a low nutritional diet principally composed of jellyfish. They are also plagued by more parasites than any other fish. The one pictured left is being attended by clouds of butterflyfish, including a wrasse or two, at a cleaning station. Their bizarre form challenges description as is evident from their various common names. 'Mola' is the Latin for millstone referring to their broadly circular, massive and grey appearance. The English term 'sunfish' relates to their habit of laying on their side at the surface, apparently sunbathing. Elsewhere in Europe, they are called moon fish and in Germany, they are also known as 'schwimmender Kopf' or 'swimming head', a description echoed by the Polish word 'samglow' meaning 'head alone'.

But, does the sunfish have a tail? Certainly, it has an elongated tail-like structure connecting the ends of the dorsal and anal fins. The origin of this structure has been subject to speculation since the early 1800s. An enticingly elegant example was proposed by the Scottish biologist, D'Arcy Wentworth Thompson. He argued that the amazing diversity of animal forms could be shown to derive from simple geometrical transformations. A prime example was that of translating the pufferfish shape to that of the sunfish. A beguiling background to this creative theory is that pufferfish and sunfish are both members of the same biological order (Tetraodoniformes) and that sunfish fry actually resemble miniature pufferfish. Sadly, despite its attraction, Thompson's theory is erroneous. The tail-like structure is no longer considered a modified caudal fin. A recent hypothesis proposes that it is actually formed by elements of the dorsal and anal fins. Furthermore, there is no trace of a tail in sunfish embryos showing that even from the earliest stages in the egg, the caudal fin is absent. This presents a new mystery—how did this curious structure of the sunfish evolve?

Location: Nusa Penida, Bali

Captivating mimicry

Among underwater photography subjects, frogfish are top favourites with divers and non-divers alike. Apart from their captivating charms, frogfish are fascinatingly efficient piscivores. As supreme ambush predators, their biology combines a unique range of deadly features. They employ two forms of deception. In one, known as aggressive mimicry, they either blend into the background or, alternatively, stand-out conspicuously by imitating a non-threatening organism. In either case, they are ignored by unsuspecting prey. For their own protection, they combine this with Batesian mimicry in which they imitate an animal unattractive to their predators. This combination is evident in the second photo on the right where a yellow and lilac frogfish is nestling against, and subtlety matching, a large tunicate which would not be of culinary interest their predators. The location, thus, perfectly positions the frogfish to launch an attack while avoiding predation itself. Another bizarre example is described in 'An odd couple' in Chapter 7. The adjacent photo on the right shows an orange frogfish being approached by a small and seemingly suicidal pufferfish which appears to risk becoming a ready meal. However, these pufferfish are quite toxic and frogfish have probably learnt to avoid them. I have observed this species entering the strike zone of frogfish before and they have always been ignored.

The dual modes of mimicry mentioned above are particularly effective for the frogfish since, although they do change colour and texture to match their surroundings, the process can take weeks. This compares with the immediate changes achieved by other fish like groupers or cephalopods such as octopus and squid. In addition to their arts of deception, frogfish have two other deadly components in their predatory armoury. The first spine of their dorsal fin has evolved to act as a fishing lure. The lure itself, known as the esca, may resemble a small fish, shrimp, or a worm. Examples of these are shown on the left with a hairy frogfish and a clown frogfish. Finally, to complete their predatory prowess, frogfish possess the fastest known strike of any vertebrate. At just 6 milliseconds, it is coupled with the ability to expand their mouths by twelve times the resting size. Another unusual feature of their anatomy is the lack of a swim bladder, which aids the buoyancy of normal fish. Combined with their less than streamlined shape, this makes them poor swimmers and they prefer to move across the reef by essentially walking on their pelvic and pectoral fins—indeed, these adapted fins do appear like feet with dainty vestigial toes. To aid their limited locomotion, frogfish do not possess gills with the typical flaps, but instead, have rear facing tubes through which they can generate a modest means of jet propulsion.

Location: Lembeh Strait, and Togian Islands, Sulawesi

Colourful confusion

Scorpionfish are ambush predators and, consequently, masters at deception. They form a large and diverse family that includes some of the world's most venomous species such as stonefish. Within this group, *Rhinopias* are distinctively exotic. Though, like other scorpionfish, *Rhinopias* are widely distributed within the Indo-Pacific region, they are rarely encountered and are considered as a 'holy grail' in underwater photography. They provide a fascinating variation on the theme of beauty and the beast and, in this context, have evolved a novel range of hunting strategies. Rather than adopting the typical approach of drably blending into their surroundings like other benthic predators, they seem to proclaim their presence by overtly standing out. A vivid example of this is pictured right, with the fish sitting in dramatic juxtaposition before a fire sea urchin. The vibrant yellow livery, set against the fiery red of the urchin, hardly seems cryptic or discrete—particularly since yellow is one of the last colours in the visible spectrum to disappear within the blue filter of the ocean. Perhaps, with its disruptive arrays of colour, shape and texture, *Rhinopias* manage to mesmerise and confuse their victims. But, we should be wary to base such impressions on our own perspective. Although these predators may seem to stand out indiscreetly, their appearance would seem to be very effective camouflage to their prey.

There is an additional practical factor involved here. Having observed *Rhinopias* feeding, they have an extra strategy as they do not primarily strike from a stationary position. They also 'walk' towards their victims in a deceptively casual way with a curious rocking motion on their pectoral fins before launching a blurringly fast strike. Potential prey are safe as long as they remain well above or to the side. So, while the *Rhinopias* has less need to be camouflaged in long profile, even to us, when viewed head on, they can appear quite cryptic. Their disruptive patterns also obscure their eyes and, for example, in the top left photo, can be really hard to detect. The various species of *Rhinopias* present a range of colours from yellow to purple and red. As shown, their large and flared pectoral fins can also feature translucent panels, which further aid deception. As illustrated in the bottom left photo, they have the typically huge gape of an ambush predator. These amazing fish are very efficient predators with a beguiling appearance. An extra irony is that their name derives from the distinctive nasal-like profile and which, although lacking a horn, makes it reminiscent of a rhinoceros—an animal not typically associated with harmonious subtlety.

Location: Ambon harbour, Maluku

Pigment patterns

Colour patterns are evident throughout the animal kingdom and many of the most intense examples are exhibited by coral reef fish. The reasons for this have long been a source wonder. Early scientific research was compromised by being unduly influenced by how we visually perceive the world. It is now established that fish do have colour vision, but much of what we know is based on two freshwater species—zebrafish and goldfish. Given the nature of their inherently colourful world, the vision of coral fish may well be quite distinct from ours, operating on a different range of the spectrum, including ultraviolet. Fish use colours and patterns in a variety of novel ways and have the ability to make quite dramatic changes through differential expansion of colour cells known as chromatophores. Such changes may be instantaneous and fleeting or of much longer duration. There is still much to understand.

Pigment patterns may function either to camouflage or visually disrupt and, conversely, to be overtly conspicuous. Examples of both are employed by frogfish as described in 'Captivating mimicry' in this chapter. Patterns and colours are used to aid communication within and between species. Rapid colour changes are often evident during mating rituals. Another example is the signalling of readiness for cleaning, as shown in 'Cleaning interrupted' in Chapter 7. Changes are also made to enhance camouflage at night as shown overleaf by the Highfin coral grouper. Patterns are used to signal aggression, as in the striped display of lionfish, and are also a common feature in symbiotic relationships such as in the Striped cleaner wrasse. The differences between juvenile fish and adults also underline the variable function of patterns. The adult Semicircle angelfish shown right presents a dramatic contrast with the two juveniles on the left, the slightly more mature lower one beginning to show the transitional phase. Although quite evident, the differences in the juvenile and adult Splendid dottybacks shown below are far more subtle. These fish are much smaller than angelfish and the similarity in their development stages may reflect a more constant lifestyle as opposed to larger fish which have different diets and predators when adults.

Location: Komodo National Marine Park, Flores

Trigger happy

There are around 40 species of triggerfish and the Titan, measuring up to 75 centimetres, is the largest. Its impressive dentition enables it to consume robustly armoured prey such as large crabs and molluscs as well as to break off chunks of coral to dine on the polyps. It also feeds on other challenging fare such as sea urchins, its back-set eyes keeping them safely away from the needle-sharp spines. Their presence on the reef is often signalled by a cloud of sand attended by a variety of smaller fish. These seek an easy meal as the Titan energetically forages into the substrate to find food. This shows its generous spirit since it does not seem to mind these opportunistic interlopers joining the feast. However, Titans have a darker side and enjoy a notorious reputation with divers. During the mating season, they aggressively defend their territory, rushing headlong towards any intruder that may pose a threat and there have been many reports from divers of Titans inflicting serious wounds. Even their approach can be quite intimidating and the natural response is to quickly retreat. Unfortunately, many divers, being unaware of the fact that the Titan's territory is cone-shaped, swim upwards which keeps them within the zone. This makes the Titan's attack even more unnerving as they can now also strike from below. Fortunately, they tend to concentrate on a diver's fins, a natural target for fish. Their genus name, *Balistoides,* means crossbow, which seems uncannily appropriate for a fish so disposed to bolting strikes.

Titans are thought by some only to be aggressive when breeding, but that is not my experience. These fish are undeniably charismatic, but they are also unpredictable. While they may often ignore an approach, one can never be sure. The moment just before impact is shown on the right. I had been calmly observing this individual when, in a sudden change of mood, it stopped its feeding and turned to launch an attack. The photo captures the split second before its teeth hit my camera housing. This demonstration seemed to satisfy its aggressive urge as it then resumed feeding. The top left photo shows a Titan at a cleaning station. Their eyes can rotate independently and one is clearly taking a backward glance at me—the fish perhaps considering whether to make an attack or continue being divested of its parasites. It is interesting that many divers are more concerned about sharks and barracuda when, in fact, Titans pose a more likely threat. The wider potential for triggerfish to inflict serious injury was underlined during WW2 after the sinking of HMS Dunedin in the South Atlantic. Of the 250 men who escaped before the ship sank, only 72 remained alive by the time they were rescued. Their three-day ordeal was described as a nightmare of torture and slow death. While they suffered attacks by sharks and barracuda, the worst of all were from hordes Black durgon triggerfish which, with their razor-sharp teeth, inflicted bites more than an inch deep. The bottom left photo shows a Red-tooth triggerfish, which is a close relative of the durgon.

Location: Pulau Dai, Forgotten Islands, South Banda Sea

Boxing clever

As their name implies, these fish are literally swimming boxes. Boxfish are also known as trunkfish and cowfish due to the impressive pair of frontal horns that some species possess as adults. Their bodies do not have scales and the streamlined sinuous flexibility we typically attribute to fish. So, although they are vertebrates, they are encased in a rigid bony carapace more reminiscent of the exoskeletons of invertebrates. This carapace is formed by fusing together small hexagonal plates and these are clearly evident in the juvenile shown in the lower left hand photo. Because of its spatial efficiency and inherent structural strength, the hexagon is a common shape in nature ranging from carbon atoms to honeycombs and basalt larva columns.

Although rather novel in vertebrates, an external rigid body in fish harks back some 250 million years to the armoured fishes of the Paleozoic era. Hexagons are not the only aspect of optimum performance of the carapace. Although the overall shape initially appears somewhat ungainly for swimming, it has been shown to be very efficient hydrodynamically, enabling the fish to finely manoeuvre in the spatially complex world of the coral reef. In fact, the boxfish geometry with its low mass and high structural strength has been adapted in the design of an energy efficient car. There is typically a trade-off between stability and manoeuvrability but the boxfish form achieves an optimum in both.

Across the species, boxfish range in size up to 45 centimetres. The Solor boxfish, shown right, is one of the smallest attaining a length of just 11 centimetres as an adult. Their hard carapace presents a robust defence to predators and, in addition, some species can secrete a potent toxin. So, it is common to encounter the adults feeding in open water. However, this protection is not effective for the bite-sized juvenile fish, which are consequently very secretive, hiding in any available cover. Two juveniles are shown in the left hand photos. At just a centimetre or so they appear like tiny jewels—if you can get close enough.

Location: Banggai Archipelago, Sulawesi and Ambon, Maluku

Midas touch

The Midas blenny, pictured on the right, is a small, engagingly photogenic but wary subject. It is very active, continuously feeding or protecting its territory. Given its common name, this individual had adopted a particularly apt home. Echoes of the famous 'Midas touch' are fittingly conveyed through the golden highlights in the translucent encrusting sponge, which adorns the entrance of its safety retreat. This fish, like many blennies, supplements its diet by grazing on algae, but its principal food is zooplankton. For a creature that dines on such tiny morsels, the males possess a pair of impressively large and sharp canines in their lower jaw, which are permanently bared. These 'pseudo teeth' enhance the display of virile prowess to deter rivals and attract mates. Despite its fearsome looking fangs, the blenny is quite shy and needs little provocation to vanish into its hole.

The other photos present examples of other species of blennies that seek refuge in the substrate of the coral reefs. Such desirable real estate may be the holes vacated by the previous tenants which embed themselves to live in the coral, thus creating deep cavities in its surface layers (see examples in 'Symbiotic embedments' in Chapter 7). The bottom left photo shows a fang blenny, which has adopted the empty tube of a plume worm which, in turn, has been encrusted by a purple sponge. This blenny leads an unpleasantly deceptive lifestyle. They mimic cleaner wrasse both in appearance and behaviour. They also have formidable canine teeth which, in contrast to those of the Midas blenny, are for more than just display. Rather than offering the service of removing external parasites, these imposters inflict further damage on unsuspecting fish attending cleaning stations by literally taking bites out of them. The process of cleaning itself can be quite uncomfortable and attendee fish will painfully recoil when a parasite is removed from a sensitive spot. To allay suspicions of their carnivorous intent, fang blennies mimic the movements of true cleaners so their bite comes as a cruel surprise. There have been reports of regular clients of a given cleaning station becoming wise to the deception and turning the tables by attacking the false cleaners.

Blennies are extensively represented in tropical and sub-tropical seas and are reported to comprise 58 genera and over 400 species worldwide. They consequently constitute an important component of small benthic reef fishes (see 'The small majority and Darwin's Paradox' in Chapter 9).

Location: Liberty Wreck, Tulamben, Bali

Horse sense

Seahorses present an engagingly visual enigma. As fish with equine heads and prehensile tails, making them more exotic than unicorns, they would be quite at home in a book of fantastic beasts. The three examples shown here are around 12 centimetres long, but their relatives, the pygmy seahorses, featured in the next section, are exquisitely tiny at just a couple of centimetres. Though their appearance is so familiar, they are one of the strangest of fish, hardly resembling anything conventionally fish-like. It is difficult to believe they share the same class with goldfish and anchovies. They belong, in turn, to a large and quite bizarre group of fishes. This embraces ghost pipefishes, trumpetfishes, cornetfishes, shrimpfishes, seamoths, sand eels and sticklebacks, including, rather appropriately for such an oddly disparate group, the paradox fish—a small freshwater stickleback found in Indonesia.

To further illustrate this group's diversity, a pair of mating seamoths is included in the bottom photo. Within this eclectic group, seahorses are assigned to the separate genus of *Hippocampus* derived from the Greek 'hippos' for horse and 'kampos' meaning sea-monster, underlining their fantastic credentials. With more distinction from convention, they have bony rings of plates rather than scales and it is the males which give birth to live young. It is an example of a 'reversed' pregnancy as the female deposits the unfertilised eggs in the male's brooding pouch, which then self-inseminates. Sadly, these amazing animals are under threat—mostly for Chinese medicine, and also, for the curio trade and aquarists—and many of their 57 listed species are on the IUCN Red List.

Location: Lembeh Strait and Ambon Harbour, Maluku.

Pygmy seahorses

The first reported specimen of a pygmy seahorse was discovered quite serendipitously. It was spotted on a sea fan that had been collected for display in an aquarium in New Caledonia by Georges Bargibant in 1969 and was named after him as *Hippocampus bargibanti*. With a total length of just 2 centimetres, they are among the smallest known vertebrates. They are very cryptic and use homochromy, a particularly effective form of mimicry in which they closely match the colour and texture of their hosts. This is enhanced by the many tubercles on their bodies, which imitate the closed coral polyps. As fish, their appearance is even more improbable than their larger counterparts. Fortunately, unlike them, they are not collected for Chinese medicine or the curio and aquarium trades, as discussed in the previous article. However, being obligate associates of a single genus of sea fans (*Muricella*) makes them very vulnerable. Despite being protected, sea fans are still illegally harvested and their environment is subject to the usual range of stressors such as climate change and pollution.

Hippocampus bargibanti have two colour variations, yellow or, more commonly, pink, as shown opposite and in the top left photo. Their prehensile tails enable them to anchor to the sea fans where they are ideally placed to feed on the zooplankton brought by the currents. Their discovery naturally stimulated a lot of interest and, in the relatively short time since then, another 7 species have been reported. The other photos on the left show two further species which are even smaller. The bottom photo shows another member of the group, the pygmy pipe-dragon, which is also known as the Thread pipefish as its body is just 1 millimetre thick. Discovered in Lembeh in 2006, it is closely related to the sea dragons of Australia.

Location: Raja Ampat, West Papua and Lembeh Strait, Sulawesi

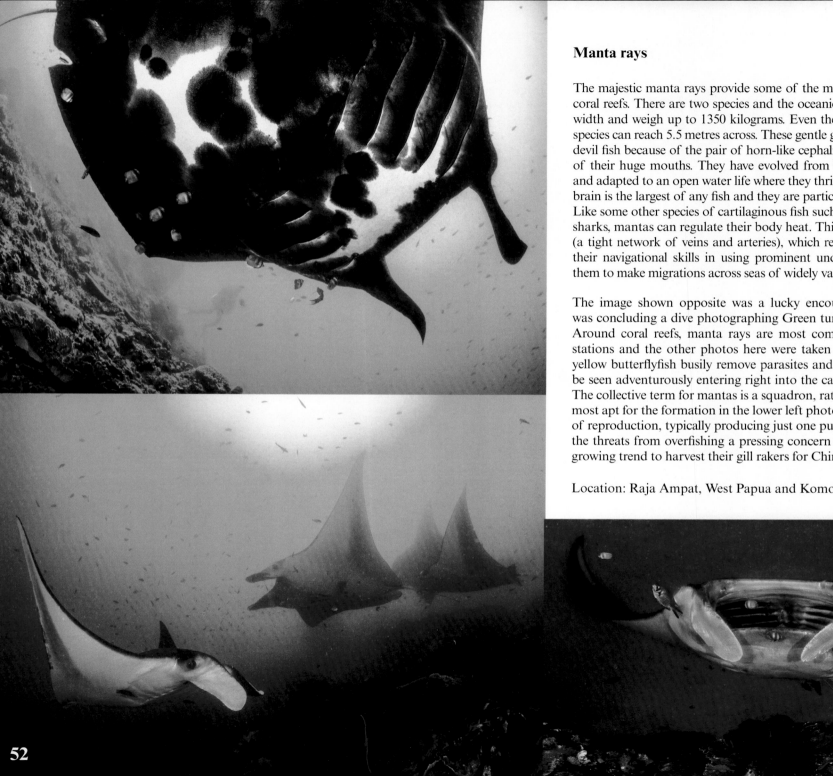

Manta rays

The majestic manta rays provide some of the most magnificent encounters on coral reefs. There are two species and the oceanic manta can attain 7 metres in width and weigh up to 1350 kilograms. Even the more frequently seen smaller species can reach 5.5 metres across. These gentle giants were originally known as devil fish because of the pair of horn-like cephalic fins projecting on either side of their huge mouths. They have evolved from the bottom dwelling stingrays and adapted to an open water life where they thrive on a diet of plankton. Their brain is the largest of any fish and they are particularly noted for their curiosity. Like some other species of cartilaginous fish such as the Mako and Great white sharks, mantas can regulate their body heat. This is achieved by a 'ret mirable' (a tight network of veins and arteries), which reduces heat loss. Coupled with their navigational skills in using prominent underwater features, this enables them to make migrations across seas of widely varying temperatures.

The image shown opposite was a lucky encounter towards sunset when I was concluding a dive photographing Green turtles (as related in Chapter 8). Around coral reefs, manta rays are most commonly seen visiting cleaning stations and the other photos here were taken at such locations. Hordes of yellow butterflyfish busily remove parasites and necrotic tissue and some can be seen adventurously entering right into the capacious mouth of the manta. The collective term for mantas is a squadron, rather than a shoal, which seems most apt for the formation in the lower left photo. Manta rays have a slow rate of reproduction, typically producing just one pup every two years. This makes the threats from overfishing a pressing concern and one compounded by the growing trend to harvest their gill rakers for Chinese medicine.

Location: Raja Ampat, West Papua and Komodo National Marine Park

The wonders of whale sharks

The whale shark is the world's largest fish. In 1987, a fishery in Taiwan recorded a female 20 metres long and weighing around 34 tonnes. To provide a sense of scale, the one shown in the right hand photo is just a juvenile of around 7 metres in length. Though whale sharks evolved some 60 million years ago, they remain shrouded in mystery. The species was not described scientifically until 1828 and relatively little is known about their ecology and life cycle. In 1995, a pregnant female, 10.6 metres long and weighing 16 tonnes, was also captured in Taiwan. While half the size of the largest recorded, it carried some 300 embryos—the highest number ever recorded for any species of shark. They ranged in length from 42 to 64 centimetres and were in varying stages of development. This confirmed that whale sharks are viviparous with the embryos developing within the uterus until ready for live birth. The gender ratio in this specimen was recorded at just over 50% with females slightly exceeding males—though the discovery raises more questions than it answers. This, along with other puzzles, is further considered in 'A double mystery' in Chapter 7.

These awesome yet gentle giants are attracted to fishing platforms, known as bagans in Indonesia, which are large wood-frame floating structures operated by local fishermen. By night, they capture huge quantities of tiny fish such as anchovies. To keep them as fresh as possible before market, the catches are held in vast nets suspended beneath the bagans. These are a magnet to whale sharks seeking an easy feast. A pair is shown in the lower left photo enjoying one of the free handouts at the bagan. This learned behaviour is encouraged by the fishermen as they view such visitations as good omens that will bring more luck to their fishing. However, they do not extend this largesse to other hopeful feeders. During my visits, these have included a sailfish and dolphins. These maintained a very respectful depth, no doubt aware that they could otherwise risk becoming a bycatch for the fishermen. A few of the dolphins are visible in the top left photo well below the whale shark where they could safely enjoy some of the titbits that sank within their reach. They were careful not to surface near the bagan, but were observed coming up to breathe well over 100 metres away.

Location: Triton Bay, West Papua and Gorontalo, Minahasa Peninsula, Sulawesi

CHAPTER 3

INVERTEBRATES

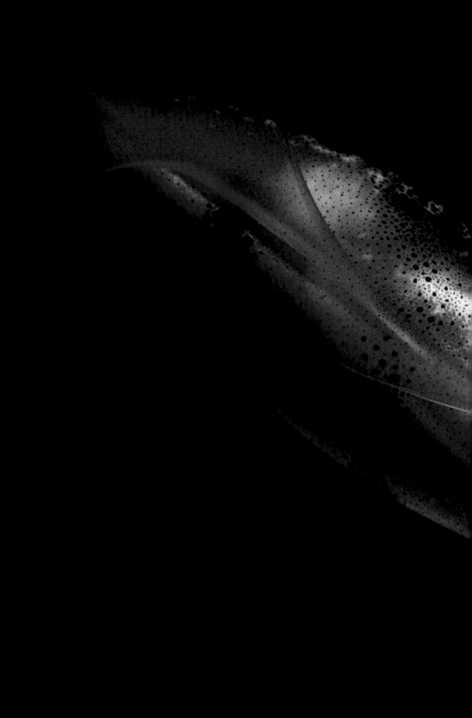

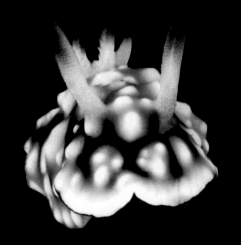

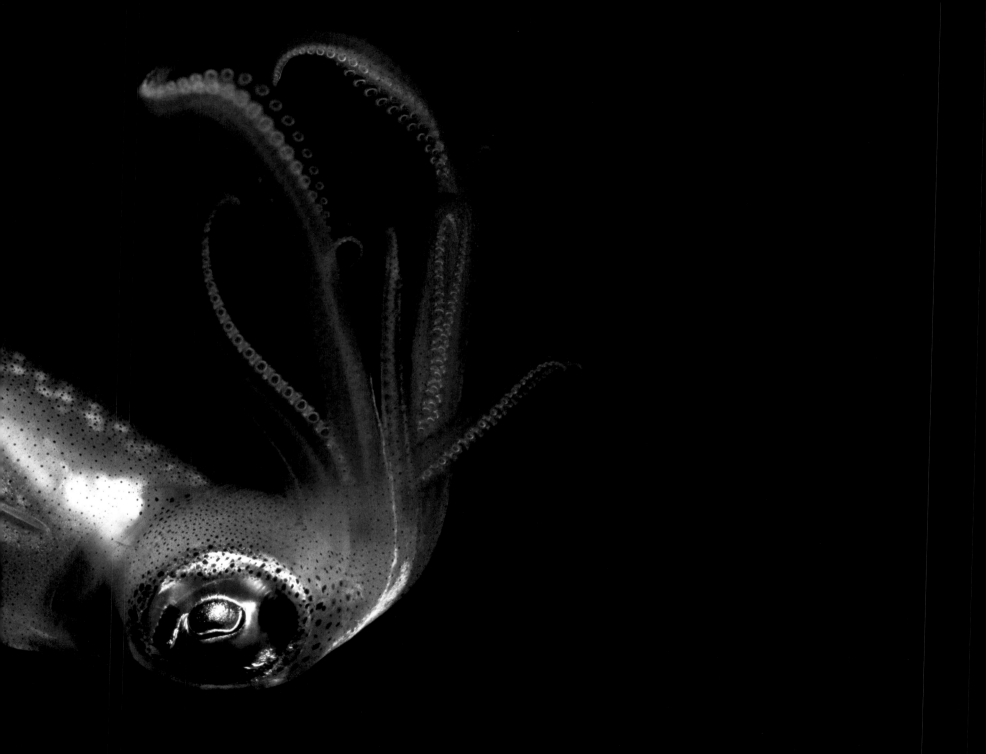

Lords of the rings

The 10 known species of Blue-ringed octopus, like other exotic creatures featured in this book, are also candidates as some of the world's most venomous creatures. The body and arm lengths range between 4–6 centimetres and 7–10 centimetres, respectively. Their venom contains tetrodotoxin, which is a thousand times more powerful than cyanide and causes respiratory arrest and heart failure. Despite their modest size, it is estimated that a single octopus has sufficient venom to kill over two dozen people. Despite such dire threats, blue-rings are highly prized as photographic subjects, especially as they are very secretive and rarely seen by divers. With this dual reputation of rarity and potency, it is amusingly ironic to discover how well known they are to the local kids. A quick chat will reveal that they regularly find blue-rings in shallow water and rock pools and, from their reports, seem quite unafraid to handle them. Whether this is just to impress through youthful prowess, it is certainly unadvisable even though blue-rings, if not harassed, are inherently docile.

During the day, they hide within the reef, emerging at night to hunt small crustaceans. Those shown right and top left illustrate some of their dramatic variety in appearance. They not only display vivid chromatic colours but also can pulse their iridescent blue rings at 3 times a second to reinforce their aposematic warning signs. The Ocellate or Mototi octopus shown below has one ring on either side. Like other cephalopods, they are masters of camouflage. This is impressively demonstrated in these photos as the octopus rapidly changed shape, texture and colour from vivid yellow to shades of muted pastel to perfectly match the substrate.

Location: Lembeh Strait and Banggai Archipelago, Sulawesi

Fatal attraction

This bizarre encounter features two very contrasting participants. A Bumblebee shrimp is pictured delicately poised at the very tip of a Textile cone shell. There are about 800 species of cone shells, which include some of the most venomous creatures in the sea and the Textile cone is one of the most potent. Around 30 accidental fatalities to humans have been attributed to these animals. Many species of cone shells are attractively patterned and can be found within easy reach in shallow water. Unfortunately, for unwary collectors, casual handling has the risk of being impaled by the shell's deadly harpoon as it responds defensively. Fortunately, for the shrimp in the present encounter, cone shells are very prey-specific and the Textile cone feeds almost exclusively on other gastropods and is also unlikely to perceive the shrimp as a threat. Both of these invertebrates are predominantly nocturnal, so it was quite a surprise to find both of them active during the day and in such intimate proximity. Perhaps the 'The Twilight Zone', which is the local name of the dive site where they were encountered, suggests a clue to such activity. The location tends to be shrouded in deep shadow as it extends beneath jetties and large moored vessels. On cloudy days, it can seem like a perpetual dusk has descended.

The omnivorous shrimp, around 2 centimetres long, was very active and presumably focussed on finding food. I followed this curious crustacean, with its rather unshrimp like appearance, as it moved across the pebbly substrate. Its progress, though energetic, could hardly be described as the buzzing of a bumblebee. Rather, its common name derives from the physical resemblance of its stripy carapace. My already heightened interest significantly increased as, directly in its path, a cone shell began to emerge from the rubble. Quite undeterred by this new arrival, the shrimp simply mounted the shell at the back, traversed its whole length, and then descended gracefully from the front of the shell. The photos to the left show the emergence of the cone shell and its full profile prior to the arrival of the shrimp.

Cone shells are the subject of extensive research both in marine biology and medical science. This relates to their complex evolution, which includes a wide family of genes and the incredible diversity of their toxins. There are at least 50,000 of these, each denoted by a gene with a specific DNA sequence. These have proved particularly valuable in creating more effective pharmaceutical drugs, for example, in controlling nerve signalling and pain relief.

Location: Ambon Harbour, Maluku

Stars and stripes

These pages feature some more weird and wonderful denizens from the world of shrimps. Tiny and cryptic, they are not commonly encountered and there has been little research into their life cycles and ecology. That shown right and top left is known as the Spiny tiger shrimp. As these photos emphasise, it certainly boasts an impressive array of spines. However, even if it does have miniature echoes of William Blake's 'fearful symmetry', the complete absence of stripes hardly conveys a convincing imitation of a tiger. Given its generous complement of spots, the name of Leopard shrimp would seem more appropriate. But, that is already taken by the associate shrimp of the Tiger anemone. Not surprisingly, since the shrimp closely matches the markings of its host, they both have stripes, as shown bottom left. So, perversely, the Leopard shrimp has stripes and the Tiger shrimp has spots. Even more improbably, the latter is also known by aquarists as the Bongo bumblebee shrimp. To add to the mix, the name Bumblebee shrimp is also applied to the species shown below, which at least does have prominent stripes.

In order to avoid the potential confusion of common names, where possible the Latin names of the species featured in this book are provided in the Photo guide in the Appendix. Whatever they are called and irrespective of whether they have spots or stripes, these fabulous creatures are stars.

Location: Ambon harbour, Maluku and Togian Islands, Central Sulawesi

Marbled exotica

Even among a fabulous variety, Marbled shrimps stand out as truly exotic crustaceans. They belong to the genus *Saron*. While they have a wide distribution globally and are extensively present in the Coral Triangle, they are not commonly encountered. They are nocturnal and even under the cover of darkness, they are notably shy and do not tend to venture beyond the protection of the reef. They are, thus, rarely seen by the majority of divers—even those who are aware of their spectacular appearance and actively seek them during night dives. However, there are always exceptions to the rule and the individual pictured on the right is a wonderfully delinquent example. In the middle of the day, we hardly expected to see a Marbled shrimp and, in any case, the sudden build-up of current made conditions difficult for any close-up photography. Luckily, my dive guide and I found a sheltered eddy. There to greet us was the shrimp boldly posing in full regalia. The elongated claw arms reveal it to be a male, as is the individual pictured top left.

The common species, *Saron marmoratus,* was described by Olivier way back in 1811 followed, nearly a century later, with *S. neglectus* by de Man in 1902. The latter name seems rather apt since there have been few studies since then even though the genus is believed to contain several unique species.

Location: Pulau Moa and Ambon Harbour, Maluku

Sitting tenants

Examples of the so-called Hairy squat lobster are pictured in the right and top left photos. Such a description seems rather prosaic for these wonderfully delicate and exotic tenants of the giant barrel sponges with which they associate. So, some prefer the fancier common name of Fairy crab, which is also somewhat more biologically appropriate since they are not really lobsters at all, but belong to a group known as Anomura crabs. They are coloured to blend into their background, a process known as homochromy in which the associates adopt the same colour as their hosts. Their extensive endowment of bristles enables them to detect the subtlest variations in water pressure, thus providing early warning of the approach of a potential predator or even a hopeful photographer. With the adults measuring just 1.5 centimetre, the labyrinthine structure of these giant sponges provides these tiny creatures with plenty of ready sanctuaries in which to retreat. Assisted by their early warning system, they are, thus, ever ready to promptly disappear at the merest indication of a threat. In practice, they avoid showy display of their visual charms by keeping safely secreted out of sight. This, of course, makes them a challenge to find and even more to photograph. Often, the only options available are rather compromised angles or awkward downward shots. Occasionally, as pictured here, an adventurous individual may be found in a more exposed location towards the outer regions of the sponge. The frontal view top left, conveys an almost feline appearance enhanced by an apparent display of bristling confidence.

Maintaining the theme of pink sponge exotica, another bizarre creature is featured in the bottom left photo. This is a sea spider. Most species measure no more than 2 centimetres and, although fairly common, they are so inconspicuous that they are invariably overlooked. As is evident in the photo, their bodies are especially diminutive—so much so that some of their organs are accommodated within their limbs. In fact, in some species, their intestines extend throughout their legs. Their tiny size does bring compensations as it eliminates the need for a normal respiratory system. Instead, with their elongated geometry providing an advantageous surface to volume ratio, they are able to breathe by diffusion through their external surface. While they are arthropods with eight legs like their familiar terrestrial counterparts, they are not arachnids. Rather, they have their own taxonomic class of Pycnogonida. Most are carnivores and typically feed on soft-bodied cnidarians such as the tiny hydroids which are shown extending their delicate filigree of tentacles from the surface of the sponge.

Location: Lembeh Strait, Sulawesi

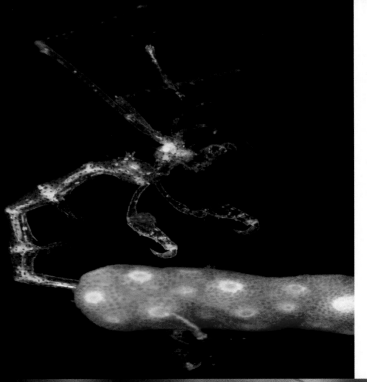

Eccentric shrimps

The underwater world of the Coral Triangle harbours a wondrous range of marine life, most of which is very small. The first challenge for the hopeful observer is discovery. The majority of these tiny creatures are very cryptic. The Hairy shrimp is a good example. Since it resembles no more than a wisp of algal matter, it is also known as the Algae shrimp. Measuring less than 6 millimetres, extreme close-up equipment is required to do them justice photographically. As shown by the individual below, their livery generally matches the drab colours of algae. The contrast is accentuated by the bright yellow polyps of the octocoral in which it sits. They are about 5 millimetres in height, and help to convey the shrimp's diminutive size. As often happens, the unexpected encounter with the shrimp pictured opposite occurred while seeking other small subjects. Gravid with eggs and arrayed in vibrant pink, this tiny jewel glows with a charming eccentricity against its muted background.

The Skeleton shrimp pictured left, while four times longer than the Hairy shrimp, is still very small. Although somewhat shrimp-like in appearance, they are not true shrimps, but are caprellids, a family of amphipods. They are omnivorous and their diet includes microscopic fare such as diatoms and protozoans. While their eccentric appearance and impressive claws make them reminiscent of the praying mantis, their reproduction is even more bizarre. Mating is only possible after the female has moulted, but before her new exoskeleton hardens. Like the praying mantis, females of some species have been observed to kill the males after mating, which, in the case of caprellids, is by injecting them with venom. The female guards the eggs until the juveniles hatch, which emerge fully formed like miniature adults. One is seen sitting on the female's antennae.

Location: Togian Islands, Central Sulawesi and Pulau Kur, East Banda Sea

Beauty and the beast

The Bobbit worm feeds into our irrational fears of what horrors may lurk in the dark depths of the oceans. Pictured right as it rears upward at night from the sea bed to seize some hapless prey, it is indeed the stuff of nightmares. With huge claws as sharp as daggers and spiked carapace, glistening in an iridescent rainbow of colours, it at once repulses with fascinates. It's a real charmer. It gained further fame as one of the stars of the BBC's Blue Planet II, but was reported to have left viewers deeply traumatised. The worm featured in the programme was a metre long, which would make any encounter daunting enough, but records show that they can grow up to 3 metres and weigh 4 kilograms. Even its name derives from dark notoriety and relates to Lorena Bobbit, who was reputed to have amputated her husband's penis in an act of revenge.

The sea bed conceals a host of creatures and many, like the fascinatingly horrific Bobbit worm, are ambush predators. But some, like Garden eels, have genuine charm. From a distance, they can appear like a waving garden of sea grass. They live in large groups and rise up from the sand during the day to feed on zooplankton as shown in the sunlight vista below. They are often the first to settle in a freshly laid bed of volcanic sand as noted at Komba (see Chapter 9). While there can be hundreds or more in a colony, they are extremely shy and can easily be missed as they readily withdraw into the safety of their burrows. With patience, they may be gradually approached enabling a portrait as shown left.

Location: Lembeh Strait, Sulawesi and Tulamben, Bali

Antennae range

The Painted spiny lobster is an impressively large crustacean attaining lengths of up to 40 centimetres. As shown opposite, its glaring red eyes signal its displeasure at the interruption to its nocturnal activities. It extends its incredibly long secondary antennae to repel my approach and holds its fan-shaped tail in readiness to retreat. Wishing to minimise my intrusion, I retreated leaving it to resume foraging for its typical fare of molluscs, crustaceans, worms and sea urchins. Unfortunately, in terms of food menus, the spiny lobster is highly rated commercially as a gourmet dish and is severely depleted in some areas. Like most crustaceans, they have two sets of antennae as can be seen in the photos. A primary pair is attached to the first segment of the carapace and another projects from the second segment. Just how long these are in the spiny lobster is shown in the top left photo taken during the day. The primary pair are biramous or branched and used to sense its environment and, of course, to find food. It always amazes me that these creatures, with such extensive appendages, are able to move around and within the complex and labyrinthine topography of coral reefs with such apparent ease.

The three species featured here indicate the range of adaptation in antennae. A slipper lobster is shown below and a Crinoid squat lobster bottom left. Despite their common names, none of these three are closely related to true lobsters and show wide variation in body form. In the spiny and slipper lobsters, this is evident from the lack of claws on their front legs, while the squat lobsters are more related to hermit crabs. Like the spiny lobster, the secondary antennae of the slipper lobster are most distinctive, but in a completely different way as they are formed into short flattened plates. The 2 centimetres long squat lobster associates with crinoid feather stars. Their long but fine antennae are barely visible though the front legs are longer than its body and equipped with substantial claws.

Location: Misool, South Raja Ampat, West Papua

Changing the guard

While many of the thousands of species of crabs have distinctive markings that disguise their outline, decorator crabs adopt a more bespoke strategy for camouflage. Some deck themselves in dead or living matter to completely disappear within the background. Others choose adornments that make them look less like crabs. This is particularly evident in the photo opposite, where the crab, dressed in a living array of white hydroids, dramatically stands out against the vivid red of the crinoid. This bold disguise is enhanced since the hydroids are armed with stinging cells and provide an additional guard from predators. These crabs have hair-like structures with tiny hooks on their bodies known as setae, which enable them to secure their chosen embellishments like Velcro. If necessary, to ensure good adherence, the crabs carefully roughen the edges of each decorative element with their mouths.

As I admired the crab's fine collection of hydroids, an intriguing transition unfolded before me. As it proceeded to move away, it revealed an ethereal impression of itself. In fact, it had just completed its moult, its new soft shell enabling it to expand to a larger size. During this process, the crab faces a double jeopardy since, along with the old shell, it must also shed its protective decoration. This must be promptly transferred from its old exoskeleton to the new and now hardening one. These photos capture the moment of completion. An example of natural recycling.

Location: Horseshoe Bay, Komodo National Marine Park, Lesser Sunda Islands

Aliens

The first night dive evokes a dream-like sensation of floating within an alien world. While not a high priority with all divers and for some, a one-time venture, it can also mark the beginning of a new romance with the coral reef. The difference from the day can far exceed the best and worst expectations. At night, the familiar rhythms and extensive vistas are replaced by an enveloping darkness which, beyond the beam of a dive torch, is punctuated with the minute flares of phosphorescent plankton. Wave your arm and you are treated to a private light show as a trail of tiny stars marks its passage. The bustling shoals of daytime fish have disappeared but may be found, dispersed one by one in wide-eyed sleep, secreted within the reef. They hide from their nocturnal predators such as moray eels and sharks while all around a vast range of other hunters appears. Many, like Golden sweepers, feast on plankton, while others, like sweetlips and squid, dine on the wide array of molluscs and other invertebrates that also emerge at night. The hard outlines of the hexacorals have acquired an unearthly softness as millions of tiny tentacles reach out to join the banquet of plankton.

Numerous examples could serve to compare the night with the day. Here, some of the less celebrated fauna of the reef are chosen. That, pictured right, appears like some alien spaceship floating across a bright cloudy sky. In fact, it is a Golden jellyfish drifting among thousands near the surface in one of the many marine lagoons in the Coral Triangle. Each lagoon is connected to the sea primarily through tiny fissures in the limestone. This isolation has led to distinctive evolution of the jellyfish between different lakes. Below is a group of Moon jellyfish. These arrived in hordes during a night dive and, as they inflict painful stings, I was very glad to be wearing a full wet suit and hood. The final picture presents an alien-like organism to complete the dark side. Commonly known as the Night sea anemone, it is one of the few to be seriously hazardous to humans with toxins comparable to those of cobras.

Location: Triton Bay West Papua and Lembeh Strait and Togian Islands, Sulawesi

Poetry in mantles

In contrast to many of the other creatures presented in this book, cowries, at least as shells, are well known throughout the world. They have been used for millennia as currency, for example, in China, India, and in many island communities—a practice which endured into the 20th century. They also became widely valued as decoration especially for jewellery and ceremonial events—particularly the Tiger cowrie as shown top left. However, cowries and their allies, the ovulids, are far less known as live animals in their full regalia resplendent their distinctive mantles as featured here. Cowries belong to the family of Cypraeidae in which they comprise around 300 species and most are herbivorous. In the majority, the mantles can be fully retracted into the shell when the animal is inactive, but will completely envelop the exterior when extended. This maintains the shell in a pristine shiny condition since the new shell layers are continuously deposited on the outside rather than inside like other gastropods. The shells have other notable features, including the formation of teeth at the aperture, as the final layer of shell covering is deposited as they become adults. This has allowed them to evolve without an operculum. Many cowries are nocturnal and the photo of the Tiger cowrie shows it foraging at night, but with its mantle partially withdrawn showing the shiny shell beneath. In passing, it may be noted that, like the Tiger shrimp discussed in 'Stars and stripes' earlier in this chapter, the pattern features spots rather than stripes providing another potential confusion with common names.

The mantles are multi-purpose, functioning as camouflage and mimicry as well as for respiration. Cowries are not known to be toxic, which raises the question of why their shells are so brilliantly coloured as they would not effectively serve as aposematic warning signs to their predators. As can be seen in the photo of an ovulid, or false cowrie, on the right, the mantle when fully extended imitates the branches of the soft coral with which it associates although to our eyes its white background does seem to stand out. Another form of mimicry is exhibited by the ovulid shown centre left in which the tubercles in the extended mantle imitate the polyps of its host. A careful marine biologist will point out that these cowries cannot count, as their fake polyps have too many tentacles. However, natural selection would indicate that their predators are not very numerate either. To illustrate other aspects of shell and mantle evolution, an opisthobranch is included bottom left. This predator has the deceptively charming common name of 'Bubble snail' and it seems to delicately glide above the substrate without actually touching it. The shell is vestigial and too small to accommodate the entire animal. They emerge from the sand at night to seek their principal diet of polychaete worms although they also hunt and consume nudibranchs.

Location: Lembeh Strait, Sulawesi

The lost chord

Ascidians are present in all seas. Their name derives from the Greek 'askidion', meaning a little wine-skin or vase. They were first mentioned by Aristotle, around 350 BC, though he considered them to be molluscs. Where they actually fit into the animal kingdom has been controversial ever since. On coral reefs, they are everywhere, but are often overlooked or ignored. This is a pity since, with over 3000 species, they have an amazing variety of forms and colours and those who study them are notably passionate. Many species are exquisitely beautiful such as the vase-like example shown on the right, which transmits a subtle translucence. To underline their proliferation, there are at least four other species evident in just this one tiny patch of reef and the main ones are identified in the Photo guide in the appendix, page 227. A colonial ascidian reaches upwards on a stalk like some other worldly tree. Although the colonial species have quite varied arrangements and structures, the basic individual form is a round sac-like body enclosed in a cellulose tunic giving rise to one of its other names of tunicate. A third name of sea squirt relates to the jets of water caused by muscle contraction when the animal is touched. Cellulose is an uncommon component in the animal kingdom and relatively indigestible. In addition to this unappetising deterrent, ascidians also produce toxins, including sulphuric acid, by absorbing heavy metals. Despite these defenses, ascidians are actively consumed by some fish, including angelfish, batfish and boxfish, and certain gastropods. In 'Naked ambition' in Chapter 7, a *Nembrotha* nudibranch is shown consuming colonial ascidians and reinforcing the predation by laying its eggs on it. Ascidians are major contributors to marine fouling and their propensity to cover substrates is evident in the left hand photo. A giant frogfish has adopted this soft bed of ascidians as an attractive perch from which to ambush its prey.

While these fascinating animals are often mistaken for sponges, they are far more complex biologically. In fact, they are our closest relatives among the invertebrates. Though sessile for most of their life, they have an initial phase as free swimming larvae known as ascidian tadpoles. These have a notochord which is a forerunner of the backbone. However, they differ dramatically from an amphibian tadpole since they have no mouths and do not feed. This phase just lasts a few hours as the tadpole soon settles and adheres itself to the substrate. Then, in a weird metamorphosis, it absorbs all of the parts that made it a chordate in the first place. Along with the notochord, these include the rudimentary eye and brain, which are no longer needed during its now sessile existence. It then promptly develops the intake and exit siphons to enable it to thrive as a filter feeder. In fact, ascidians are able to regenerate their whole bodies from just a small segment of a blood vessel. It is an amazing life cycle.

Location: East Nusa Tenggara, Lesser Sunda Islands, Alor

CHAPTER 4

PREDATION

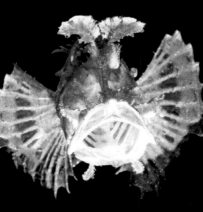

The first link

The fundamental importance of plankton cannot be overstated. Phytoplankton form the first link in the vast food chain that vitally sustains all marine life. This chain extends from phytoplankton to zooplankton, to the 'small majority' of the myriad species of small fish and other tiny creatures, to larger species such as fusiliers, and so on through to the majestic manta rays and the largest fish of all, the whale sharks. This and the following pages illustrate examples of this vast range of fish species that feed on plankton.

In dense shoals, fusiliers pictured right, intersect and overlap in a rather chaotic aggregation above a red barrel sponge. These large shoals provide dynamic encounters on Indonesian coral reefs. Here, they are not displaying the synchronized harmony of schooling, as evidenced by their multi-directional orientations, but are simply gathering to inspect me as a diver. This curiosity is common, but soon satisfied as these very active fish return to their constant feeding as they seek their diet of zooplankton. They inhabit the mid-water zones above the reef and in terms of global biomass, fusiliers constitute one of the most important groups of coral reef fishes. The photo was taken on a glorious sunny day in spectacular visibility at a depth of 20 metres on a submerged sea mount. As emphasised in Chapter 2, fish are innately curious and, unlike birds in a forest, will often approach quite closely—especially if one is suitably calm and quiet. On this occasion, their initial investigation was short and sweet. However, knowing the likelihood of at least a second pass, I maintained my position. Suddenly, I was completely surrounded by fusiliers arriving from all directions as they swarmed around me. Such visitations are almost guaranteed if one has the added benefit of being able to wait in serene silence using a rebreather.

Predictably, there is a multitude of zooplankton specialists on coral reefs. Anthias are a prime example and two of the 90 or so species in the genus *Pseudanthias* are shown on the left. Both photos feature males which are typically distinguished from the females by their more elaborate physical form and often colour. Large harems of females are maintained by a dominant male though if it disappears, a leading female will change sex to replace it. Described as jewels of the reef, these small brilliantly coloured fish are actually close relations of groupers. Like fusiliers, anthias also feed in mid-water and may be seen rising in dense clouds from their refuges in the reef as shown in the following two pages.

Location: Kai Islands, East Banda Sea, Maluku

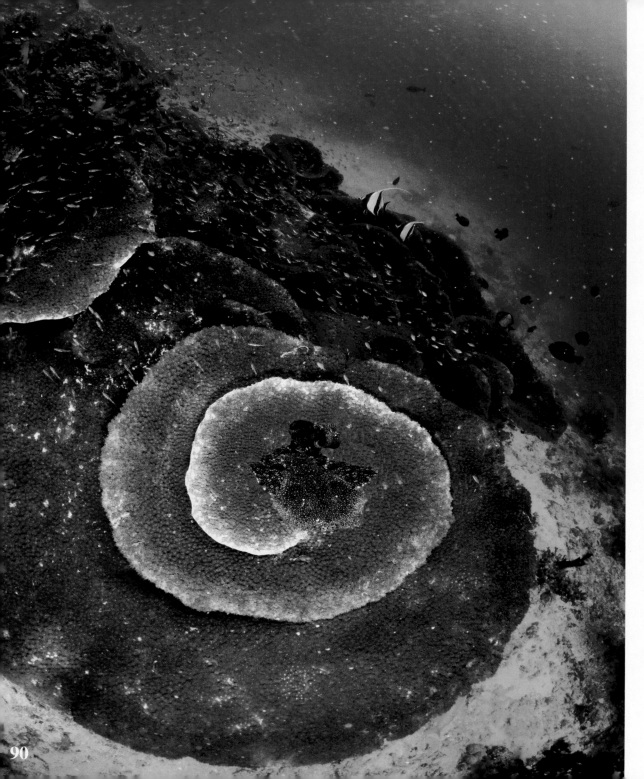

Flying carpets

The popular image of a shark is of a sleek, grey, free swimming piscivore—the ultimate predator. However, there is a whole range of less active sharks that have a sedentary lifestyle, spending much of their time just lying around. These include the carpet sharks of which there are a dozen known species—their very name conveying relaxed lifestyle. They are masters of camouflage and blend deceptively into the background. As they can attain lengths exceeding 3 metres, the larger ones, as shown opposite, can effectively create their own seascape. Such full diurnal exposure is fairly unusual since they are night-time hunters and typically hide in caves or under ledges during the day. Clearly though, if they feel sufficiently concealed, even in plain sight, they may be encountered in the open. This is also evident in the other photo, which features a very young juvenile cradled in a spiral of hard coral well up on the open reef.

These sharks are also known as wobbegongs which is an Australian Aboriginal word meaning 'shaggy beard'. The species featured here is a Tasselled wobbegong, which delicately enhances its camouflage with an elaborate lacy fringe. These filigree tassels, which extend the full perimeter of its large mouth, are far more than just adornment. They are multi-purpose and, apart from softening its outline, they contain an abundance of sense organs which enable the shark to detect even minor movements of potential prey. And, when it comes to feeding, these 'carpets' fly into action. They are consummate ambush predators. The wobbegong has a huge head which is wider than its length and has a correspondingly capacious mouth. Its jaws are powerful and armed with dagger-like fangs. It patiently waits for its prey to venture within range—either simply oblivious of its presence or attracted by those deceptive fronds of tassels. The shark has even evolved a further strategy in which it waves its alluringly curved tail, thus tempting its prey within the strike zone. That attitude of the tail seems predisposed since it is clearly evident in both the adult and the juvenile shown here. With explosive energy, the strike is stunningly fast—the sudden gape of the huge mouth creating a strong suction to draw its prey inwards.

Location: Misool, South Raja Ampat

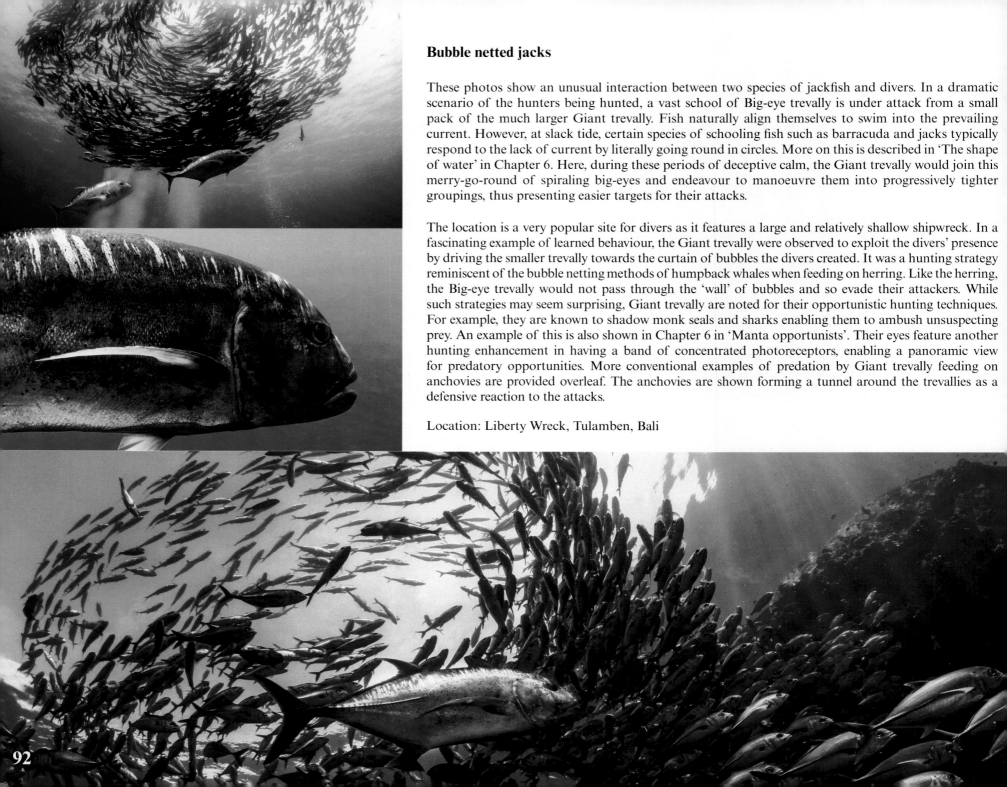

Bubble netted jacks

These photos show an unusual interaction between two species of jackfish and divers. In a dramatic scenario of the hunters being hunted, a vast school of Big-eye trevally is under attack from a small pack of the much larger Giant trevally. Fish naturally align themselves to swim into the prevailing current. However, at slack tide, certain species of schooling fish such as barracuda and jacks typically respond to the lack of current by literally going round in circles. More on this is described in 'The shape of water' in Chapter 6. Here, during these periods of deceptive calm, the Giant trevally would join this merry-go-round of spiraling big-eyes and endeavour to manoeuvre them into progressively tighter groupings, thus presenting easier targets for their attacks.

The location is a very popular site for divers as it features a large and relatively shallow shipwreck. In a fascinating example of learned behaviour, the Giant trevally were observed to exploit the divers' presence by driving the smaller trevally towards the curtain of bubbles the divers created. It was a hunting strategy reminiscent of the bubble netting methods of humpback whales when feeding on herring. Like the herring, the Big-eye trevally would not pass through the 'wall' of bubbles and so evade their attackers. While such strategies may seem surprising, Giant trevally are noted for their opportunistic hunting techniques. For example, they are known to shadow monk seals and sharks enabling them to ambush unsuspecting prey. An example of this is also shown in Chapter 6 in 'Manta opportunists'. Their eyes feature another hunting enhancement in having a band of concentrated photoreceptors, enabling a panoramic view for predatory opportunities. More conventional examples of predation by Giant trevally feeding on anchovies are provided overleaf. The anchovies are shown forming a tunnel around the trevallies as a defensive reaction to the attacks.

Location: Liberty Wreck, Tulamben, Bali

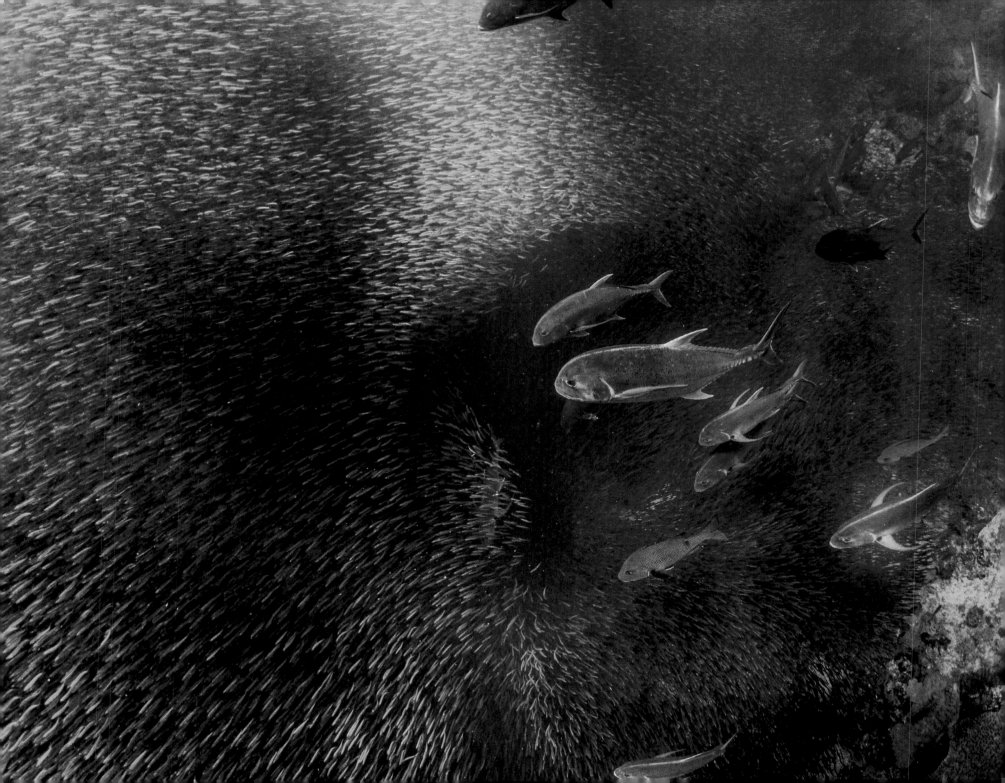

The art of ambush

The coral reef is alive with a vast army of ambush predators ranging from mantis shrimp and Bobbit worms to frogfish and wobbegong sharks. Scorpionfish are some of the most abundant though, because of their ability to achieve highly effective camouflage, their presence may pass unobserved. This was the case with the large scorpionfish discretely positioned in the foreground of the right-hand photo. I was completely absorbed with the bigger picture, seeking an upward shot of the shoal of fusiliers passing above in an attractive arc. Using a fisheye lens brought me within touching distance of the scorpionfish before I realised that it was there. One is made painfully aware of their presence if they touched accidentally. With the limited vision afforded by a dive light, this is a particular risk at night especially, as these fish are active nocturnally. They are defensively armed with sharp venomous spines that can deliver a potent sting. The severity of this can vary from moderate irritation to severe pain and, in rare cases, to fatalities.

Scorpionfish are voracious predators and though they are nocturnal hunters, dining on fish and crustaceans will also feed during the day. This was evident with the three examples shown here. These fish were not just camouflaged simply biding their time until nightfall. Like other ambush specialists, they patiently lay in wait for long periods, remaining motionless on and within the cover of the reef. Should prey venture close enough, the strike is delivered with deadly speed. Scorpionfish are very artful at blending into a wide range of backgrounds. The individual pictured top left matches the sponge from which it launches its attacks. In fact, the photo captures the success of one such foray since, on close inspection, three small fish are still visible in its mouth. The attack was over so quickly that, although I was intently observing the fish, I was barely aware that it had moved. As a further example of their creative stealth, the scorpionfish in the lower left photo has buried its body within a soft coral, disguising its presence by imitating both the colour and texture of its borrowed shroud.

Location: South Pulau Mansuar, Raja Ampat

Goby dessert

Perhaps the title is taking liberties with the dining menu of the lizardfish as, given its size, the unlucky goby is probably the main course of the day. The photo on the right, taken just seconds after the strike, depicts the final desperate struggle of the goby moments before it is swallowed alive. Held by the needle-sharp inward facing teeth, which are also present on the tongue, the goby has no chance of escape. Lizardfish owe their name to their reptilian appearance conveyed by their markings, slender pointed head and wide capacious mouth. Some 75 species are known worldwide though only a few are common around coral reefs. The Reef lizardfish shown here attains a length of nearly 30 centimetres.

Lizardfish are diurnal hunters preying on small fish and invertebrates. While not achieving the total camouflage of scorpionfish described in the previous article, they do blend in well with their surroundings. They are also ambush predators but, rather than the lightning-fast close-range strikes of scorpionfish and frogfish, they use their streamlined torpedo-like bodies to dart forward from longer range. While normally solitary, they are sometimes seen in pairs and, as shown, will often sit in the open on the reef. They also favour sandy patches and may bury themselves to enhance their disguise for surprise attacks by adopting the tactics of the stargazer family, as shown by the individual pictured below.

Location: Lembeh Strait, Sulawesi and Ambon, Maluku

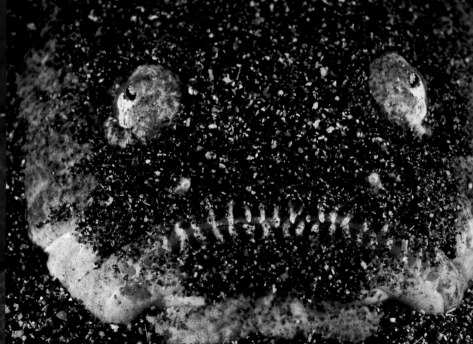

Sneak attack

Trumpetfish do not hunt in groups like jacks or barracuda or have the sudden explosive strike of ambush predators that lay in wait camouflaged on the substrate of coral reefs. Instead, these slender 80 centimetres long fish exploit covert tactics enabling them to launch surprise attacks from very unexpected places. They are voracious predators of small fish which they suck in pipette—like through their extended jaws. Their elongated shape is ideally suited to glide discretely amongst sea fans or through the arms of feather stars, as shown in the bottom photo, and where they can locate sheltering prey. Trumpetfish are also well-informed on the dietary habits of other fish. They have developed a talent for shadowing subjects that their own prey do not associate with danger. Thus, if you wish to know which fish are not piscivores, then observing which species hunting trumpetfish choose to shadow, can provide a useful guide.

The three shadowed species presented here are all benthic feeders of crustaceans or molluscs. For optimum effect, the selected decoy should be large enough for the trumpetfish to hide behind. Two such examples are shown right and bottom left, where the trumpetfish shadows an adult male Napoleon wrasse and tails a Blue spotted stingray. Displaying creative flair, if large subjects are unavailable, this artful predator will also hide amongst a shoal as shown with the Yellow goatfish in the centre left. With no suitable fish available, I have even seen a trumpetfish hanging vertically against an anchor chain and mimicking its appearance. Fish would seem smart enough to borrow a good idea and a jackfish is shown in the top left photo, adopting the same predatory advantage by shadowing a Napoleon wrasse.

Location: Pulau Moa, South Banda Sea

Pharyngeal jaws retracted

Pharyngeal jaws extended

(X-rays courtesy of Dr. Rita Mehta, UCSC)

Twice bitten—the moray's inner eel

A giant moray eel with its gaping mouth full of sharp recurved teeth is an engaging sight—even if only its head protrudes from its lair. As an apex predator at the top of its food chain, it has no natural enemies. Morays spend much of their time secreted deep within the narrow serpentine spaces of coral reefs and have evolved to exploit such constricted living accommodation. The most obvious aspect is their greatly elongated shape. Other adaptations include the loss of the pectoral and pelvic fins, the absence of scales and the ability to produce substantial quantities of body mucus for lubrication. However, their extreme shape has also created challenges not suffered by other large piscivores, which exploit speed of attack through a combination of ambush and inward suction to catch and swallow their prey. The moray is not adept at such tactics and instead tends to hunt nocturnally preying upon sedentary fish sleeping within the reef such as wrasse and triggerfish. Large prey provide very attractive and energy efficient meals. However, even when captured, they present the moray with quite a consuming problem. Because of its shape, the moray has a relatively small mouth and limited oral space. Moreover, its narrow throat prevents it from producing the strong suction that other piscivores spontaneously generate with the sudden opening of their jaws and throat expansion as they rush forward to grab and swallow their prey.

To address this challenge, the moray has evolved a unique solution. It has an inner eel! Amazingly, it is armed with a second set of moveable jaws set deep in its throat. These pharyngeal jaws can leap forward to grab prey held in the oral jaws and then drag it down the throat, thus overcoming the moray's limited ability to swallow large items. There is nothing else quite like this in the real world, although a sensationally similar example occurs in science fiction with the monsters of the film 'Alien'. Given the long-established knowledge of these pharyngeal jaws, it may seem surprising that their true role was not fully appreciated until quite recently. This was because the action of these secondary jaws is so fast that it is impossible to observe under normal conditions. It was only identified through studying high definition slow motion film and was first reported in 2007.

Location: Ambon, Maluku

Fishing fans

While the countless numbers of plankton feeders are dominated by fish, ranging from tiny blennies and gobies to manta rays and whale sharks, many species of invertebrates also depend on this bounty. These include filter feeding molluscs, barnacles and certain species of porcelain crabs. The three species of porcelain crabs and the barnacle featured here achieve this using fine nets of fishing fans. During feeding, these gossamer fans are busy in constant movement and one needs to match the rhythm to obtain an effective image. While the barnacle has a single fan, porcelain crabs deploy a total of six nets, three each on each of highly modified pairs of front legs. These are portrayed in all their fragile delicacy in the photo on the right.

All four of these invertebrates associate with cnidarians—the crab top left with an anemone, the other crabs with soft corals, and the barnacle shown below embeds itself into hard coral. Despite its large claws, the anemone crab is relatively peaceful and basically uses them to defend its territory which, between other crabs and anemonefish, can be quite keenly contested. Porcelain crabs are not true crabs, but are more related to squat lobsters. They have three pairs of walking legs as opposed to four in a true crab. This difference is clearly evident in the bottom left photo as the crab moves along the spine of a sea pen.

Location: Horseshoe Bay, South Rinca, Komodo, Flores

Legless in Ambon

There are around 4500 species of shrimps worldwide, of which two thirds are marine and most are scavengers. Like crabs, they belong to the order of decapods, which literally means 'ten footed'. Curiously, while quite a range of crab species are decorators, as described elsewhere in this book, there are no known examples of decorator shrimps—though many share similar environments, diets and predators. Most shrimps have muted colours and discreet markings. However, *Saron* (as described in 'Marbled exotica' in Chapter 3) and the Harlequin shrimp featured here are spectacular exceptions. Although avidly sought as subjects, neither of these incredibly beautiful species are often seen. They are very secretive and generally hide within the reef or beneath rubble or rocks. *Saron*, while also being nocturnal, are extremely shy. Harlequins are quite different and my first eventual encounter brought both delight and surprise. Apart from their improbable, almost 'Alice in Wonderland' appearance, the confidence they displayed, even in the middle of the day, was really unexpected. The harlequin's prime source of nutrition is the sea star which, as starkly portrayed in the photos, they slowly eat alive. They form a male and female pair, working together to hunt and manage their giant-sized prey, and to defend their territory from other pairs. These diminutive shrimps, though no more than five centimetres long, are easily able to immobilise their prey by turning it over onto its back. They then drag their victim off to their lair within the reef. In this grim sea star prison, they leisurely dine on the tube feet and other soft tissues, methodically moving along one of the legs. The hapless sea star is, thus, held prisoner at a gristly banquet in which it is the sole menu. Some species do use a means of escape by shedding the doomed appendage. Amazingly, harlequins are reported to encourage this by partially amputating a leg close to the central disk. One isolated leg is much easier to manage and can last the pair a couple of weeks. Consequently, harlequins are often found with just a single leg of a sea star, as shown on the right. Shedding a leg is a very effective if extreme survival strategy since, as long as they are not fatally injured, sea stars are able to regenerate. In fact, sea stars are occasionally encountered simultaneously regrowing multiple legs as shown by the Multi-pore sea star, bottom left, which is achieving the feat of regenerating four.

Harlequin shrimp prey on several species of sea stars though the Blue sea star, top left, and the Red sea star of the *Formia* genus, pictured right, seem particular favourites. If their sea star larder is temporarily depleted, harlequins are known to supplement their diet with the tube feet of sea urchins. With their wondrously patterned livery, they do not attempt to camouflage. This along with that apparent bold confidence strongly indicates that the colouring is aposematic, warning potential predators that they are toxic or, at least, have a most unpleasant taste.

Location: Ambon Harbour, Maluku and Lembeh Strait, Sulawesi

Stunning weapons

There are around 450 species of mantis shrimp and two of those commonly encountered in the Coral Triangle are presented here. They are not actually shrimps, but belong to a group known as stomatopods. Neither are they related to the praying mantis from which their common name derives. Rather, it is because they share highly modified forelimbs that are employed as deadly attack weapons. There are two main types of this strike force—those with clubs that smash their victims and others that impale with spear-like pincers. The Giant mantis shown opposite is of the latter kind. This species can reach lengths up to 25 centimetres and live in vertical burrows formed in the sand. As the burrow is in a soft substrate, it is easily enlarged to accommodate their increased size after each moult. The cast-off remnant of the lower half of one of their spearing limbs is visible in the bottom right of the photo. Measuring 7 centimetres, it usefully conveys a sense of scale. The mantis typically stay flush with the burrow entrance, scanning their environment with some of the most advanced eyesight in the animal world. The large and prominent eyes are mounted on stalks and each can swivel separately. While our eyes which have three types of photoreceptors, theirs have over a dozen and can detect a much wider spectrum of wavelengths, including ultraviolet. Each eye has over 10,000 of these photoreceptors arranged in a configuration that enables trinocular vision. This, rather amazingly, means that they can see in 3D with each eye independently. How they process such complex visual information is intriguing. They mainly feed on fish, shrimps and squid, but will take any potential meal that passes above their burrow. I have tracked a squid at dusk, itself hunting close to the sand, only to see it suddenly vanish as it is violently seized from below and dragged into the lair of a Giant mantis. The strike is incredibly fast and delivered with such force that their victims are sometimes cut in half.

The Peacock mantis shrimp shown left, though a little smaller at 18 centimetres, comes in more spectacular colours. It shares the same formidable repertoire of offensive weaponry except that its forelimbs are calcified to form clubs. These are used to smash its prey with such force that they can easily crack open the shells of quite large crustaceans and molluscs. The acceleration achieved compares with that of a 0.22 calibre bullet. It is so rapid that the water literally boils in the partial vacuum formed around the club and a flash of steam has been captured on high definition slow motion film. This, in turn, causes cavitation bubbles which produce a secondary stunning force from the shockwave when the bubbles collapse. The victim is, thus, exposed to two sequential impacts. The second one from the shockwave can be enough to stun prey if the first strike does not fully connect. To avoid damage to the mantis itself, the forelimbs have shock absorbers. Unlike, the Giant mantis, Peacock mantis often leave their burrows and may be seen out on the reef. The lower photo shows a brooding female. The clutch of eggs is so large that she is unable to hunt effectively. Mantis shrimp are reported to form monogamous pairs with the male providing for both of them during the brooding period.

Location: Ambon, Maluku and Lembeh Strait, Sulawesi

Night watch

The reef undergoes an almost surreal transformation from day to night. Entering the water just before dusk, one can actually witness this transition. Shoals of daytime fish rush by, moved by the dual anxiety of finding a safe haven for the night and avoiding predators whose shadowy forms dissolve invisibly in the fading light. Then, like an enveloping cloak, peace briefly descends. Soon, in ever-increasing numbers, the creatures of the night slowly emerge. The wiry arms of brittle stars weave around sponges, basket starfish spread their arms, an army of crustaceans leave their daytime retreats as do the uncomfortably numerous sea urchins. Soldierfish, with the large eyes of a nocturnal hunter as shown opposite, emerge to feed on crustaceans and small fish. A hermit crab, top left, treads a delicate path across the accommodating polyps of a hexacoral which, surprisingly, do not retract. Now fully extended, each polyp reaches out to ensnare and feed on the night bloom of plankton.

One of the strangest actors in this nightly theatre is the pearlfish. As shown in the bottom left, it is a translucent fish with an eel-like slenderness. They have a predilection for entering orifices from which they emerge at night to feed on small fish and shrimp. Their hosts include oysters where they have occasionally been found encased in mother-of-pearl and from which their name derives. The sea cucumber is its preferred abode which they enter through the anus. One is pictured below adorned with a garland of juvenile brittle stars. Added to the pearlfish, it looks a distinctly uncomfortable affliction. Sea cucumbers must have high tolerance levels.

Location: Weda, Halmahera, North Maluku

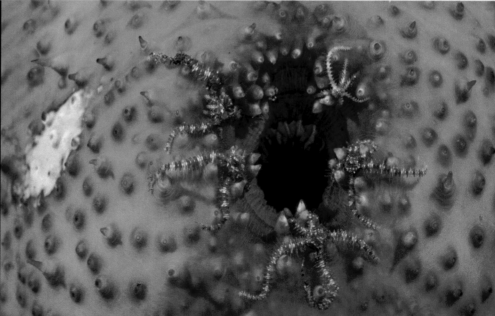

Strange intelligence

Squid, like other cephalopods, are fascinating creatures. They have taken a totally different evolutionary path to vertebrates. This is immediately conveyed by their bizarre appearance and anatomy. In fact, they have no skeleton at all and the only really hard body part they possess is their beak, which is used to kill and dismember their prey. Just in itself, the beak is an amazing piece of bio-engineering. It is made of chitin, a natural derivative of glucose, and yet the business end is incredibly sharp, harder and more resistant than most metals. Squid may appear to us like some alien life form from science fiction—an effect uncannily conveyed in the top left photo. They possess an intelligence which, though distinctly different, in some ways matches our own and in other respects, exceeds our abilities. Of course, intelligence is an elusive quality to define, but they certainly have large brains, which share a complexity comparable to ours. They feature folded lobes and specific centres for processing different sources of information such as tactile and visual stimuli. This is highlighted by their startling capacity to control a dynamic chromatic palette. With this, they are able to stage their own light shows which, especially at night, are magical spectacles. Their outer surface is crowded with chromatophores. These are tiny organs containing pigment sacs, which they can rapidly expand and contract. The colour changes are produced so fast that bands of different hues can be seen to move over their bodies like waves. To complement this, squid have excellent vision with eyes that employ advanced technology. While we focus by changing the shape of our lenses, squid achieve this with moveable lenses more like those of a camera.

The species most frequently seen around reefs in the Coral Triangle is the Bigfin reef squid shown here. They may reach up to 40 centimetres, but most encountered are less than half this size. Being on many predator's menu from sharks and other large fish to members of their own species, they are very wary of approach. During the day, they are typically seen in groups of similar-sized individuals—possibly to reduce the risk from their tendency towards cannibalism. The best opportunities for close observation during daytime occur when they are busy depositing their festoons of egg sacs on coral heads or moorings. Then, the instinct to procreate may overcome their normal wariness. Night dives offer improved chances when the squid are active, usually singly or in pairs, hunting crustaceans and small fish. But, one must still be very discrete. Overly bright dive lights or loud exhalations tend to result in their rapid disappearance leaving just a cloud of ink. With patience, they may be approached quite closely and, as a dive light can itself attract their prey, they will sometimes approach to exploit this extra hunting resource. The lower left photo shows an embryo developing in the egg sac. As a further mark of cephalopod intelligence, studies have shown that they have already learnt to identify predators before they hatch. As their parents die soon after mating and are, thus, not around to protect their progeny, this early learning phase is particularly useful.

Location: Triton Bay, West Papua and Ambon Harbour, Maluku

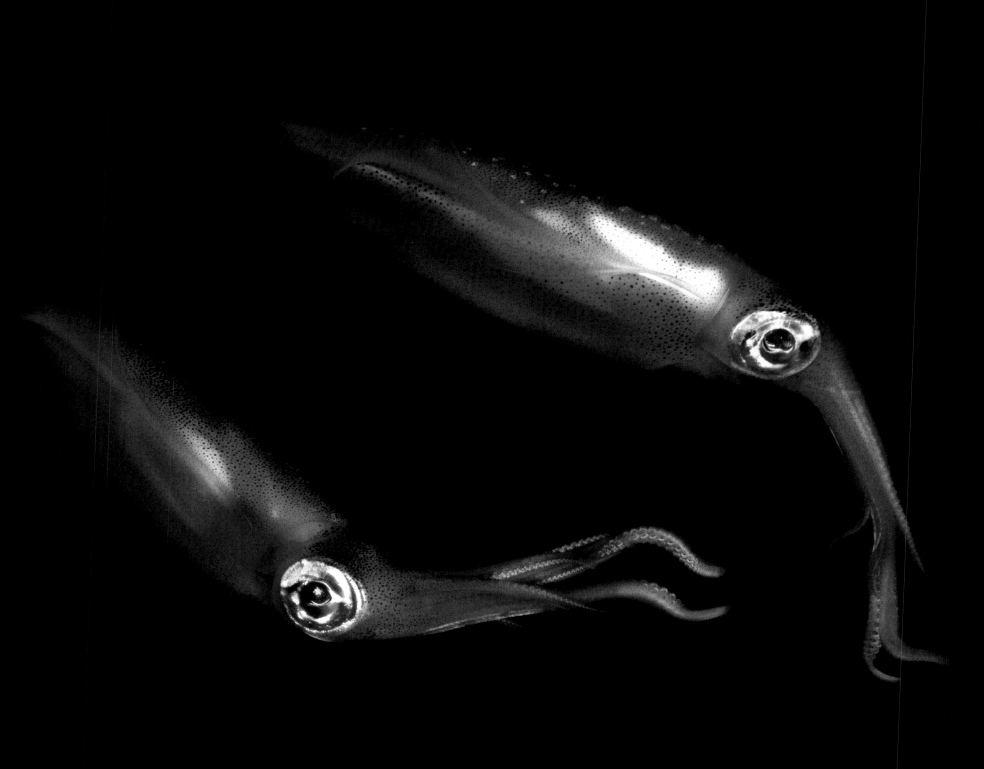

CHAPTER 5
REPRODUCTION

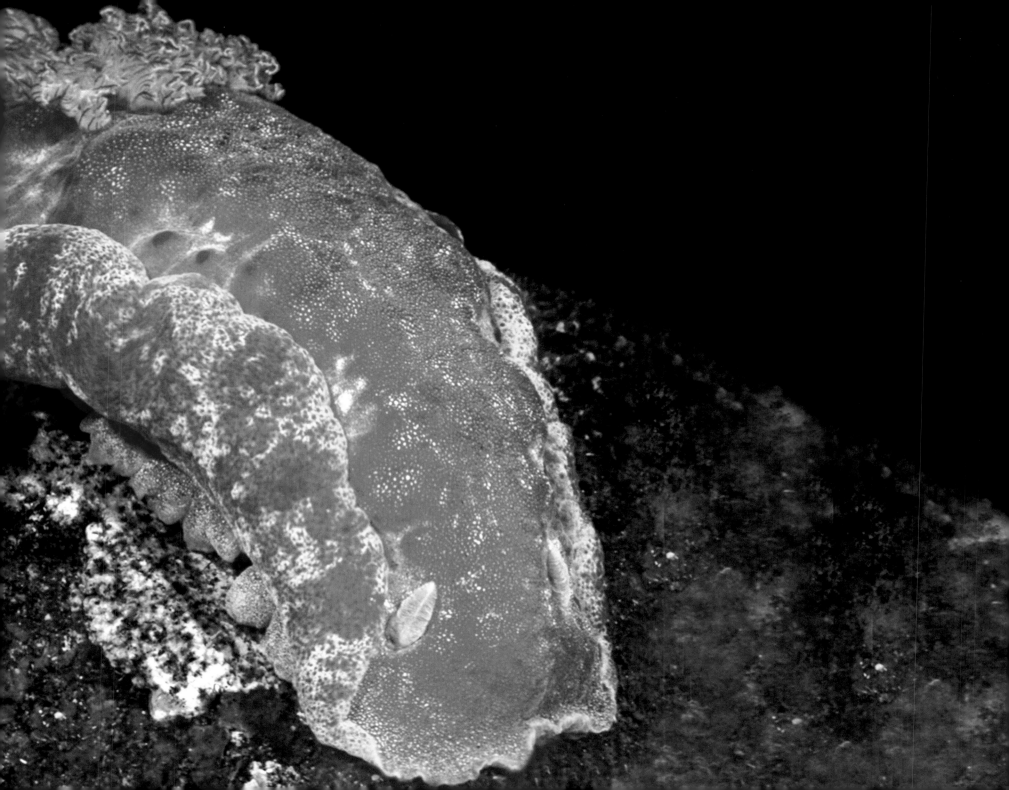

Naked ambition

While nudibranchs are found throughout the world's oceans, the Coral Triangle is home to the greatest number of species. Whether this diversity, in common with other biota, principally arises within the Coral Triangle, or is mainly due to accumulation from adjacent regions, continues to be a hot debate in biogeography. Probably, and most optimistically for the future of coral reefs, such rich variety is significantly enhanced from a combination of both.

The word nudibranch, which is derived from the Latin word 'nudus' and Greek word 'brankhia', means naked lungs, as these soft-bodied gastropods have no external shells, which they shed in the rudimentary form at the end of the larval stage. Their other and prosaic name of sea slug hardly does justice for some of nature's most extravagant and colourful creations. Apart from their spectacular decoration, these fascinating animals have a wide range of senses, including sight, taste, smell and touch, if only in a fairly primitive form. They have short life cycles; so, when it comes to reproduction, they are impressively creative. Upon reaching maturity, they have just a few months in which to mate. While they are all functioning hermaphrodites, each having both male and female organs, they need to pair as they do not self-fertilise—apart from in one known species, a sapsucking sea snail. So, this sharing of sexual resource is fundamental and occasionally, they may indulge in a ménage à trois, as shown opposite. Some species also practise protandry, which involves a young mature male and much larger female, as shown top left. Studies have shown that, for some species, mating only occurs in this way. The lower left photo shows a pair of equally sized nudibranchs about to dock. In the photo below, a nudibranch, having mated, is laying its yellow ribbon of eggs on the stalk of a colonial ascidian upon which it is simultaneously feeding, accompanied by its associate Emperor shrimp. A mass spawning event is shown overleaf and the chapter page (115) features a Spanish dancer with its bright pink garland of freshly deposited eggs.

Location: Wetar, South Banda Sea, and Lembeh Strait, Sulawesi

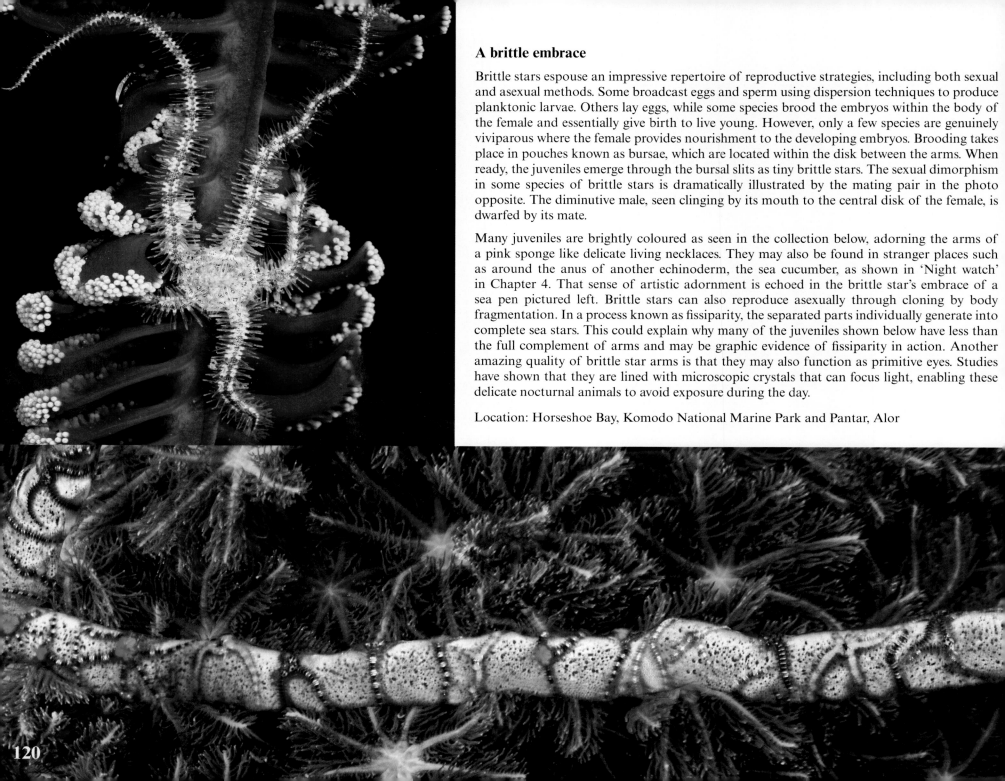

A brittle embrace

Brittle stars espouse an impressive repertoire of reproductive strategies, including both sexual and asexual methods. Some broadcast eggs and sperm using dispersion techniques to produce planktonic larvae. Others lay eggs, while some species brood the embryos within the body of the female and essentially give birth to live young. However, only a few species are genuinely viviparous where the female provides nourishment to the developing embryos. Brooding takes place in pouches known as bursae, which are located within the disk between the arms. When ready, the juveniles emerge through the bursal slits as tiny brittle stars. The sexual dimorphism in some species of brittle stars is dramatically illustrated by the mating pair in the photo opposite. The diminutive male, seen clinging by its mouth to the central disk of the female, is dwarfed by its mate.

Many juveniles are brightly coloured as seen in the collection below, adorning the arms of a pink sponge like delicate living necklaces. They may also be found in stranger places such as around the anus of another echinoderm, the sea cucumber, as shown in 'Night watch' in Chapter 4. That sense of artistic adornment is echoed in the brittle star's embrace of a sea pen pictured left. Brittle stars can also reproduce asexually through cloning by body fragmentation. In a process known as fissiparity, the separated parts individually generate into complete sea stars. This could explain why many of the juveniles shown below have less than the full complement of arms and may be graphic evidence of fissiparity in action. Another amazing quality of brittle star arms is that they may also function as primitive eyes. Studies have shown that they are lined with microscopic crystals that can focus light, enabling these delicate nocturnal animals to avoid exposure during the day.

Location: Horseshoe Bay, Komodo National Marine Park and Pantar, Alor

Anemone invasion

The parasitic Tiger anemone exploits sea fans as living scaffolds, which it overgrows to establish its own colonies—from which it acquires its other common name of Gorgonian wrapper. It, thus, steals prime real estate and is perfectly placed in the current to capture its own planktonic food. In turn, it provides a new niche environment for various crustaceans, including the so-called Leopard shrimp. These, like many other species of shrimp, live in male/female pairs and their cryptic livery closely matches that of the anemone. However, as discussed in 'Stars and stripes' in Chapter 3, the common names here present an example where their limitations become confusing. The anemone, as shown on the right, has multiplied so extensively that almost all of the original sea fan have been enveloped. The anemone deploys two methods of invasion. Apart from planktonic larval dispersion, it is also able to spread pelagically by individual polyps detaching and disbursing in the current. The top photo on the left shows one freely drifting prior to settling to colonise a new gorgonian, as indicated in the middle photo. The bottom photo illustrates how the anemone gradually advances along a gorgonian branch. The associate goby is set to lose its natural host as it is overgrown by the anemone. The fish may, however, be more stressed by its own affliction of copepod parasites seen attached just behind its head. A flourishing example of the gorgonian is shown in 'Hide and seek' in Chapter 6.

With climate change and the other pressures generated from the ever-increasing demands of our growing populations, the potential threats caused by invasive species have become more critical. At least, the anemone described here is a natural invader. Far more worrying are the invasions caused by human intervention such as the examples of the Banggai cardinalfish in Chapter 5 and, more particularly, the lionfish in Chapter 9.

Location: Togian Islands, Gulf of Tomini, Central Sulawesi

Endangered yet invasive

Banggai cardinalfish are endemic to the Banggai archipelago in Central Sulawesi. However, they now suffer the unique distinction of being both endangered and invasive. Like other cardinalfish, they are mouth brooders. The top left close-up shows a brooding fish with the maturing eggs visible in its mouth. Compared to other species of cardinalfish, they nurture much fewer fry, but to a more developed stage. This adds further distinction of a breeding sequence that lacks a pelagic dispersal phase. Coupled with their geographic isolation, this effectively restricts their natural distribution to the shallow reefs around the 30 or so islands that make up the archipelago. It is, therefore, quite surprising to encounter these attractive fish in other parts of the Coral Triangle and significantly distant from their natural home. Populations have been reported in Lembeh, Luwuk and Bali. It has been presumed that a few escaped during transportation for the aquarium trade, but other reports suggest that the introductions may have been more deliberate. The first individuals of these suspected fugitives were reported in Lembeh a decade or so ago. In the relatively short interval since then, they have drastically multiplied and are now present there in unnaturally high concentrations. Their dominating presence enveloping a closed anemone, shown on the right, presents graphic evidence. The resident anemonefish are usually very aggressive, tirelessly protecting their host anemone from any intruding fish. But here, they appear overwhelmed by the hordes of cardinalfish. In Lembeh, it is now common to see the various species of anemones infested with Banggai in all sizes, from tiny juveniles to adults. These invaders swim closely around and shelter within their tentacles; so, it is also evident that, like the anemonefish, they have also developed some immunity to the stinging tentacles of the anemone. The other photos show juvenile Banggai associating with fire sea urchins and even attempting to establish a relationship with a pair of mating *Janolus* nudibranchs.

Why the Banggai have thrived so successfully on the dark volcanic sand slopes of Lembeh Strait is puzzling. Such terrain differs significantly from their endemic environment having far less coral and a wider temperature range. Their natural predators such as scorpionfish and moray eels are widely prevalent. In Lembeh, other species of cardinalfish have not multiplied out of control and the populations of Banggai in their natural location are not seen in the aggregations encountered in Lembeh. Whatever the reasons for the disparity, it is sobering to note how easy it is for our actions to significantly upset the natural balance. It is a real irony that Banggai cardinal fish are now listed as a threatened species in their original habitat.

Location: Lembeh, North Sulawesi

(Photo courtesy of Andrew Podzorski)

Plagues and pufferfish

Pufferfish are not readily associated with plagues, but the incidence related here raises such a question. This event was encountered in the Togian Islands in April 2019. Having also visited the region in November 2006 and 2007, and witnessed similar examples on both occasions, I had assumed that it was a relatively common occurrence. At times, the shoals of these small fish were so dense that they darkened the whole reef. As shown on the right, their invasive presence seemed to be causing some consternation to the territorial anemonefish. If normal, these events were examples of nature's profligacy, since they featured mass mortality with dead and dying fish present in considerable numbers. The reasons for this are unclear. These fish are usually found in pairs, keeping very close to the reef and not high in the water column, as shown here. They are also noted to be territorial. These aspects alone make the mass aggregations puzzling. Such over-production could cause a temporary food shortage, but these fish are omnivores; so, a wide range of nutrients should be available. The apparent regularity of these events and the numbers of fish involved tend to indicate no underlying negative trend.

Our world has always been afflicted by plagues, particularly infectious diseases. Their dramatic potential has been tragically demonstrated by repeated outbreaks and most potently by the recent coronavirus pandemic. While many plagues are natural, there is growing evidence that their severity may be exacerbated by anthropogenic factors such those on the climate and the way we exploit animals for food. Plagues in the marine environment are also well known but, like the terrestrial ones, are not fully understood. Even those we impose directly like the plastics that plague our oceans have unintended and unexpectedly critical consequences. Some plagues arise naturally like those of the infamous crown-of-thorns sea star which voraciously consume corals. Even so, our actions may play a critical role through the uncontrolled collection of the Giant triton snail, the key predator of the Crown-of-thorns. These formidable sea stars also take advantage of stressed corals as highlighted in 'Destructive fishing methods' in Chapter 9. In 1983, a devastating plague caused the mass mortality of the *Diadema* sea urchin in the Caribbean. Their demise led, in turn, to widespread coral death. With the sea urchins no longer present to consume the overgrowing algae, it bloomed like a plague to infest the coral. Caribbean corals have also been devastated by the 'white plague' thought to be caused by a virus. Again, with the increased proliferation of this plague following coral bleaching, a link to climate change is suspected. Breakouts of a wasting disease have annihilated more than 20 species of sea stars from Mexico to Alaska. While these epidemics occur naturally, the latest research indicates that global warming may significantly increase the sea stars' vulnerability. Here, we are again confronted with the potential devastation of whole ecosystems, in this case that of kelp forests, as the sea star is the main predator of the sea urchins which consume them. Everything is connected and the loss of a key species can lead to a domino effect, sending shock waves through whole ecosystems. There are stark lessons here to be learnt for the protection of the Coral Triangle, and indeed, reefs worldwide.

Location: Togian Islands, Sulawesi

Pipe dreams

The evocatively named Ornate ghost pipefish is a member of a wonderfully bizarre and improbably delightful Order of fishes called gasterosteiformes. As noted in 'Horse sense', these include the paradox fish, seamoths and their closest relative the seahorse—which could be described as a curled pipefish. Pipefish are the most specialised yet prolifically represented in this group with over 225 species and 50 genera worldwide—easily exceeding those in any other family. By comparison, seahorses are contained within the single genus *Hippocampus*. The greatest diversity of pipefish, as with many other biota, is found in the Coral Triangle.

It is fairly well known that the reproductive organs of seahorses have evolved to create the odd scenario of male pregnancy. The males, on receiving the unfertilised eggs from the female, effectively self-inseminate within their brooding pouch and eventually give birth to live young. Perhaps not surprisingly with such a disparate and eccentric group, some species of pipefish do things differently—even departing from their closest relatives. While seahorse sexes are similar in size, the Ornate ghost pipefish females are significantly larger than the males and their reproduction takes a more conventional path. With these pipefish, it is the female that incubates the eggs in the capacious brooding pouch between her pelvic fins. In the photo on the right, the eggs can be seen at an early stage showing as tiny spheres within the female's pouch. A batch at a much later stage is shown on the left in which the eyes of the developing embryos are now evident. Even so, the pouch is distinctly different to that of the male seahorse, which remains fully closed until the young finally emerge. By contrast, the pouch of the female Ornate ghost pipefish, being formed by the pelvic fins, is always open and allows continual ventilation of the eggs. The Slender pipefish pictured below further emphasises the range of variation in form and function, as it is the male again which self-inseminates and broods the eggs. In contrast to the male seahorse, the pouch is an extended slit and the glistening red eggs can be seen bulging from it. The physical limitations of the elongated shape impose limits on the male brooding capacity. The females often produce more eggs than the males can manage and they occasionally resort to embryo cannibalism to lighten the load.

Location: Ambon Harbour, Maluku

Replicating rainbows

Mandarinfish are considered by many to be the most beautiful of marine fishes. These exquisite creatures, no more than 6 centimetres long, have the suitably evocative Latin name of *Synchiropus splendidus*. If anything, their bright, psychedelic livery is more than just splendid. It is particularly dramatic when set against the drab backgrounds of the rubble and coral debris that they inhabit. Like the other species of dragonets, mandarinfish have no scales but are covered in a thick toxic mucus. This indicates that their bold colours and patterns serve as aposematic warnings. This fish is further distinguished by producing a truly blue pigment. This colour in most vertebrates is not a pigment, but is created by interference patterns through layers of crystals. The only other vertebrate known to have true blue pigment is the Blue poison dart frog. The feathers of Blue jays, for example, only appear to be blue because of light diffraction.

Mandarinfish share their predilection for modest habitats with another contender for the beauty crown—that of the Flasher wrasse, whose brightly coloured and visually startling regalia is featured next in this chapter. They also favour the low light at dusk to gather for their mating rituals. However, compared to the frenetic antics of the mating wrasse, the courtship and spawning behaviour of mandarinfish is a quite sedate affair. The couples rise slowly together from the rubble as if partners in an elegant dance, delicately fluttering their pectoral fins. At times, in the pervading gloom, they seem to emerge like some mystical apparition as if being drawn upward on invisible threads. At the very peak of their rise, they hover, entwined for just a second or two, to complete the act of spawning. This brief moment is captured in the photo opposite as a tiny cloud of sperm and eggs may be seen rising between them. Such spectacular events are a magnet to underwater photographers though achieving successful portraits is quite a challenge. In the fading light, the fleeting appearance of the fish makes it impossible to see and focus in time without artificial illumination. However, the indiscreet use of bright lights will completely inhibit the shy mandarinfish and they will remain, if somewhat sexually frustrated, within the rubble. This disappointment is shared by the hopeful photographer since, once disturbed, the mandarinfish show is generally over for the night.

The males tenaciously defend their territories and may guard a harem of many females. Sometimes, a particularly successful male will rise with several females simultaneously during the spawning ritual. The acquisitive males are naturally bolder than the females and, if one waits quietly, they may approach quite closely. This confidence is indicated in the close-up of a male in the top left photo, which was taken from very close range. Mandarinfish show conspicuous sexual dimorphism with the males being considerably larger than the females, as illustrated by the mating pairs in the photos.

Location: Lembeh Strait, Sulawesi

Flasher frenetics

In our globally connected and well-travelled planet, Triton Bay in West Papua stands out as remote, idyllic and relatively untouched. It is breathtakingly beautiful above and below the surface. So, it may seem strange to conclude a visit to such a paradise by quietly hovering over a sandy rubble slope in fading light well away from the main coral reef. However, this is the favoured habitat of some of the most colourful and visually spectacular dramatic species of fish—the Flasher wrasse. Wrasse, in general, are noted for their vibrantly colourful livery, but the flashers take visual exuberance to a new level. The photos capture the males in full mating mode in the fleeting moment when they hold their dorsal fin erect with its dramatic array of spines. The startling comparison between a displaying male and the modestly patterned female is shown on the right.

These theatrical subjects present an ultimate challenge in macro photography. They are quite small, reaching a maximum length of around 5 centimetres. During the mating activity, they move extremely fast along erratic and quite unpredictable paths. The males pause to display for just a second or two. Twenty species of this genus have been recorded and there is often more than one species involved in the same mating ritual. This has generated the highest recorded frequencies of hybrids in any species of fish. The hybridisation in the males is visually evident in the variation of colour, pattern and the number of dorsal spines. An idea of the range of such variations is illustrated here. The mating activity becomes increasingly frenetic towards sunset and concludes, in the dying light, with a dazzling display of sexual frenzy.

Location: Triton Bay, West Papua

CHAPTER 6

BEHAVIOUR

The shape of water

While the strength of currents may be visually evident at the surface, when underwater, one may need other indicators. In this respect, the shapes adopted by schooling fish can provide useful insights. The circular formation of the barracuda, as shown on the right, is commonly likened to a tornado. These violent creations of thunderstorms spin with destructive energy and velocity. Ironically, this pattern of schooling fish arises from quite the opposite conditions. Fish will orientate head-on into the prevailing current. Consequently, it is the very absence of current between tides, providing no preferential direction, which leads fish to adopt these circular formations. It is, therefore, a popular misnomer to associate these silent circles with the violent weather of high-speed whirlwinds. So, while notably dramatic in aspect, the barracuda are effectively resting in neutral. The only common factor with a tornado, apart from the similar geometry, is perhaps the stillness at the centre. The fish are certainly not in feeding mode and it is possible to safely swim into the middle to be enveloped within a curiously gentle merry-go-round of these renowned and voracious predators.

These types of circular formations are typically encountered in schooling predatory fish such as barracuda and jacks. The latter may be seen swimming in this orientation in 'Bubble netted jacks' in Chapter 4, though in this case, it is also partly a defensive reaction to the predatory intentions of the larger trevallies. The even more tornado-like appearance of the school of scad in the bottom left photo is wholly generated by defensive tactics as in the 'bait ball' formations of sardines and other small prey fish when under attack.

Returning to the barracuda, as soon as the tide starts to turn and the current builds, the fish will promptly begin to unwind and spiral away to orientate to the current, as demonstrated in the top left photo. It is important for divers to be aware of changing conditions, especially in the many areas of restricted underwater topography where strong currents can rapidly generate. It is easy to become completely immersed in the experience of close encounters with such majestic fish. But, they can also convey important information. Their changing formations signal the transience of the moment and the need for divers to respond to changing conditions.

Location: Misool, South Raja Ampat

Manta opportunists

Some piscivores adopt a hunting strategy by shadowing a non-threatening surrogate, enabling themselves to spring surprise attacks on unsuspecting prey. The trumpetfish is a prime exponent of such tactics, as illustrated in 'Sneak attack' in Chapter 4. Large planktivores such as manta rays attract a range of opportunists, as shown here. Their size makes them attractive mobile launch pads for substantial piscivores such as the Giant trevally. A trio of these is seen trailing a manta ray in the right hand photo, with all four participants tracing an artistic arc through the water column. It appeared that one or more of the trevally was trying to dock into a good hunting position close on the back of the manta, as shown in the example top left. However, on this occasion, the manta was apparently not in an accommodating mood as it energetically continued with its spiralling paths obliging the pursuing trevally to become partners in an aquatic ballet. Since the location was close to a cleaning station, and the trevally would happily consume cleaner fish, I wondered whether the manta was determined to first rid itself of this inconvenient attendance. If so, the tactic proved successful and the trevally eventually abandoned their pursuit, though it was wonderful to watch while it lasted. Other manta opportunists include the yellow striped pilot fish in the lower left photo and, of course, the divers who get front seats to observe cleaning stations frequented by manta rays, as shown below.

Location: Misool and Palau Mansuar, Raja Ampat

Juvenile behaviour

Some species of batfish have evolved a special talent for disguise and deception during their juvenile stages. The adults prominently gather in open water above the reefs, making no attempt at concealment. In marked contrast, the juveniles featured here, while also frequenting exposed locations, actually manage to hide in plain sight. They achieve this by mimicking dead leaves drifting at the surface. They even suspend themselves by placing the tops of their dorsal fins in the surface tension, enabling themselves to also emulate the motion of a drifting leaf. It seems an adventurous strategy as such exposure, being silhouetted at the surface, should make them potentially very vulnerable. There are predatory fish below and hungry opportunist birds above. However, evolution has delivered an illusionist's master trick. These juveniles successfully exploit a double benefit by being safely cryptic in highly exposed locations while simultaneously being able to enjoy a meal. Thus, disguised at the surface, they are well placed to calmly pick at algal growths on floating material. At least two species of batfish, both shown here, adopt this mimicking behaviour when juveniles. Other species of batfish keep well away from the surface as juveniles and have developed quite different survival strategies. For example, juveniles of the Pinnate or Longfin batfish mimic a toxic flatworm, as described in 'Safe haven' in Chapter 1.

The batfish featured here were encountered beneath a jetty where the shaded, diffused light aided their disappearing act. Nearly all of those shown, including on the following two pages, are of a single species—the Round batfish. The sequence of photos graphically captures the range of colours and patterns which they adopt while at the surface. They soften their stripes and turn to faded yellow or brown to match dead leaves. The second fish in the row of four to the right is seen to most impressively achieve the leaf-like illusion. The top left photo shows another four juveniles, complete with a real drifting leaf, which emphasises their mimicry. The lower left photo shows two Round batfish juveniles followed by a young juvenile Blunthead batfish. It does not seem to make a very convincing impression of a drifting leaf, at least to our eyes, but it would naturally seek the security of being part of a group. More pattern variations are provided overleaf and they also illustrate how these fish will adopt the mimicry even under bright conditions in very shallow water.

Location: North Raja Ampat, West Papua

Hide and seek

Filefish are closely related to triggerfish though they have more laterally compressed bodies. They are also, unlike their relatives, able to make relatively quick colour changes to match their backgrounds. In some species, like the Seagrass filefish pictured to the right, the bodies of adults and older juveniles feature tiny projections or tassels. These further enhance their camouflage, and the fish is clearly able to adapt its disguise to more than just shades of green as its name might otherwise imply.

There are over 100 species of filefish. They are particularly well represented in Australia and quite a few are common in the Coral Triangle. The largest attain lengths up to 75 centimetres, but many are quite small. Adult Seagrass filefish are less than 12 centimetres long and those of the Puffer filefish on this page attain a length of only 5 centimetres. Filefish are omnivorous and feed on a wide range of benthic invertebrates, including ascidians, bryozoans, echinoids and sponges. The juveniles are very adept at hiding and, aided by their flat profile, can promptly disappear gliding between coral polyps or the plumes of tube worms matching colours as they go. The photo of the tiny juvenile below, measuring less than 5 millimetres, was taken as it briefly transferred between worms.

Location: Wetar, South Banda Sea

Dancing classes

In their endeavours to avoid predation, fish have evolved an impressive range of defences. They deploy various forms of camouflage and deception, using different types of mimicry, to either blend into or stand out from the background. Physical defences include sharp and venomous spines or armour plating, while others rely on safety in numbers. Many of these methods are described elsewhere in this book. Yet another tactic is employed by the juveniles of certain species fish—notably those of sweetlips, two of which are featured here. These are the Harlequin sweetlips, pictured near right and below, and the Striped sweetlips to the left. The juvenile on the far right is that of the Humpback grouper. While displaying a variety of patterns, including spots and stripes, they all adopt a similar tactic to deter and presumably confuse predators. If approached, they will suddenly adopt a most eccentric swimming motion, holding their head down and waving their bodies in a curious undulating motion. It has been suggested that they may be mimicking toxic flatworms. While this may conceivably apply to the Striped sweetlips juvenile, the other two do not have any obvious resemblance to flatworms found in the Coral Triangle. At least the Pinnate batfish juvenile, as discussed in 'Safe haven' in Chapter 1, does resemble the flatworm that it mimics. So, while this strange dancing motion is clearly successful, being adopted by various species, the reasons remain unexplained.

Location: Pantar, South Banda Sea

Shape shifters

Cephalopods are well known for their intelligence and incredible ability to change their appearance through rapid and dramatic variations in colour, shape and texture. In the Coral Triangle, a species of octopus has taken these skills to a new level. The eponymously named Mimic octopus, discovered in Sulawesi in 2000, displays a mesmerising repertoire of shifting shapes. This enhanced capability to confuse and deceive is known as dynamic mimicry and may be deployed both offensively and defensively. Mimic octopuses make their homes in burrows in the soft sand and sediment typically found in the dark volcanic deposits of Lembeh Strait, and also, in many other similar environments in the Coral Triangle. Although octopuses are generally nocturnal, Mimic octopuses may occasionally be encountered at large during daytime hunting small fish and crustaceans or seeking a mate, which I have also witnessed in broad daylight. With sharp observation, they may more often be spotted from a distance surveying their surroundings as shown on the top left. However, such long-range detection is an expertise of dive guides and the octopuses are likely to retreat well before achieving any proximity. Diurnal forays are not typical behaviour of other species of octopuses since they would risk running the gauntlet of predatory fish. So, what is the Mimic's secret? Extensive observation has revealed them to employ a bewildering range of shapes and patterns in what appears to be dynamic imitations of other sea creatures. Since this mimicry is essentially visual, it would support the contention of significant daytime activity since such disguises would not serve at night. Judging by the confidence exhibited by Mimic octopuses during the day, their dynamic endeavours must be sufficiently convincing both to their prey and potential predators.

Three of the most commonly cited disguises adopted by Mimic octopuses are illustrated here. That pictured right has been proposed to represent either a lionfish or a toxic anemone. The centre left could perhaps be construed to imitate a sea snake. These two examples are either toxic or venomous creatures which have aposematic markings to warn off predators. The Mimic too would seem to display such warning colouration so, while not yet scientifically verified, the hypothesis of dynamic mimicry may seem to hold. The third example bottom left shows a Mimic in flight mode and is said to resemble a flounder or sole. This analogy seems more tenuous, however. Although some species of flatfish are common in the same environments, they rely on camouflage rather than standing out with warning markings. Moreover, only one of the many genera is reported to have powerful toxins; so, at best, it would be a very selective choice of mimicry. It may just be an efficient swimming shape and I have seen other species of long-arm octopus, which do not mimic, adopting it. Likewise, the shape presumed to be sea snake mimicry is also a swimming mode practised by other octopuses. In any case, the Mimic's protean talents to rapidly shape shift between displays while simultaneously flashing different colour patterns is likely to confuse the most determined of predators.

Location: Weda Bay, North Maluku and Lembeh Strait, Sulawesi

Mercury rising

The mercurial and aptly named Flamboyant cuttlefish even manages to stand out from other cephalopods—a field well crowded with exhibitionists. This small cuttlefish, measuring less than 8 centimetres, is a master of camouflage and can completely merge into the sandy substrate so well that it is virtually impossible to discern its outline. However, as described below, it also has quite an extrovert side. Most cephalopods tend to be nocturnal hunters. It may, therefore, seem surprising that this diminutive creature may be seen confidently parading across the sand during the day. The word parading is used advisedly since it can display the most colourful of uniforms as it literally marches across the substrate. In a form of locomotion unique to this genera, it uses two of its lower arms and the rear parts of its mantle to amble along if with a rather eccentric gate. However, this wonderful little creature is far from pedestrian in nature. Its confidence, so evident during its daytime forays, derives from its toxins, which have a potency not matched in other cuttlefish. It advertises this deadly threat by displaying some of the most dramatically aposematic patterns in the animal world. As shown in the top left photo, it produces a living palette of vibrant colours varying from violet and crimson through to orange and yellow. If approached when in its drab livery of camouflage fatigues, it will suddenly discard them to reveal its spectacular colour show. One may watch entranced as wave after wave of changing colours wash across its body, during which it extends its arms in a variety of eccentric postures. Its default mode when not under threat is not always one of total camouflage, as I have seen them from afar dressed in their most colourful attire. They are also impressive opportunists. On occasion, despite my close presence to photograph them in full aposematic plumage, they have calmly continued to feed. They would suddenly shoot out their feeding tentacles to grab some attractive morsel such as a shrimp that had inadvertently wandered within range. The lower left photo shows these feeding arms in action though, in this case, with the cuttlefish in more subdued attire.

The photo on the right shows some very unusual behaviour—even for a Flamboyant cuttlefish. They are not good swimmers; so, I was surprised to see this one lift away from the substrate. Then, to my growing amazement, it continued upwards towards the surface. Fascinated, I followed its slow rise in the water column. At around a depth of 5 metres, it stopped and hovered and we both hung together for a while observing each other suspended in the blue. Conscious of my rather eccentric dive profile, I made the first move and descended back to the reef, leaving the cuttlefish in its high-rise location. Back on the sand, after checking my equipment, I was surprised and delighted to find the cuttlefish just behind me. It had followed me back down where it calmly resumed its idiosyncratic progress over the sand.

Location: Lembeh Strait, Sulawesi

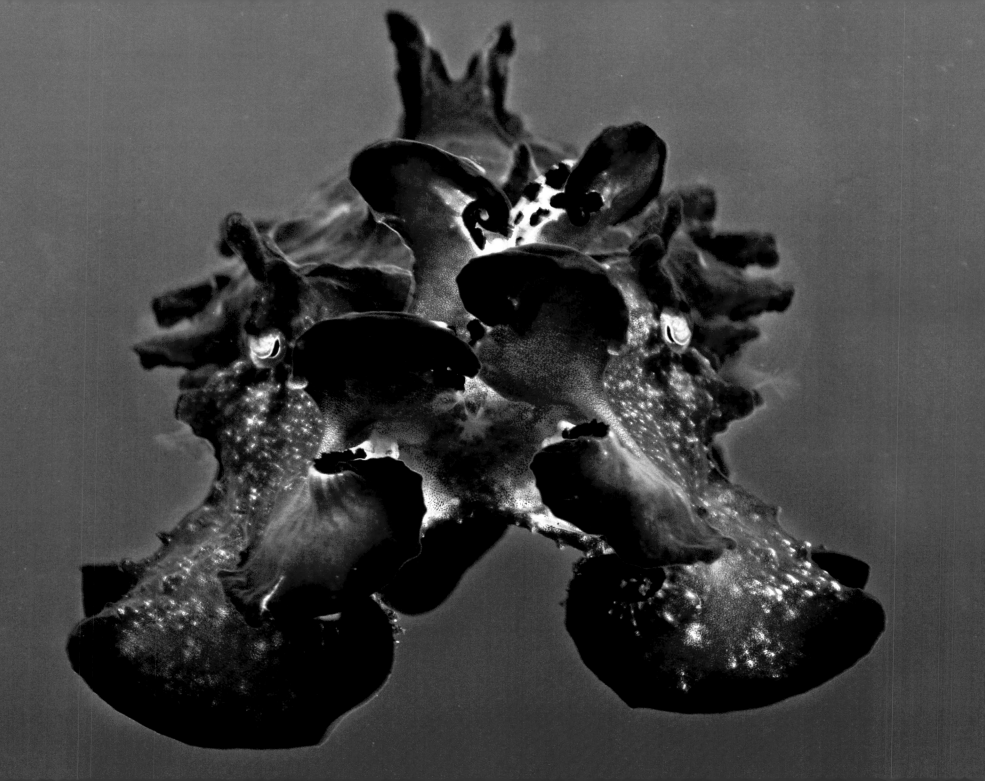

Coconut octopus

The use of tools was once considered to be the exclusive preserve of humans as a key marker of superior intelligence. The distinction has been refuted by a range of discoveries in other animals. Early examples were observed in primates—famously by Jane Goodall in the 1960s in her pioneering work with chimpanzees. Many other instances have since been reported in a growing variety of vertebrates, including mammals, birds and even fish such as the Orange dotted trunkfish in Australia. The subject is somewhat contested by which behavioural criteria define tool use, especially with invertebrates. A key factor is the ability to select specific items for future use. This rules out, for example, the placing of material to disguise or protect the entrances to lairs, the attachment of debris for camouflage, or the use of shells by hermit crabs.

If one was to predict where tool use would be encountered with marine invertebrates, cephalopods would be the prime candidates. And so, it proved in 2009 when a team of marine biologists reported such behaviour in the Veined octopus—now fondly called the Coconut octopus. As shown in the series of photos presented here, the octopus uses discarded coconut shells to make shelters. The case here was made on the basis that half a shell could be more of a hindrance than a help and that the octopus would actively seek compatible halves to complete its shelter. They were also observed to carry the halves around for use when needed. I have witnessed the similar use of small bivalve shells by juvenile octopus. It is, therefore, conceivable that this practice has been adapted by the adults with the growing availability of suitably large man-made artefacts. This behaviour also extends beyond coconut shells, as shown in 'See-through security' overleaf.

Location: Seririt, North Coast, Bali

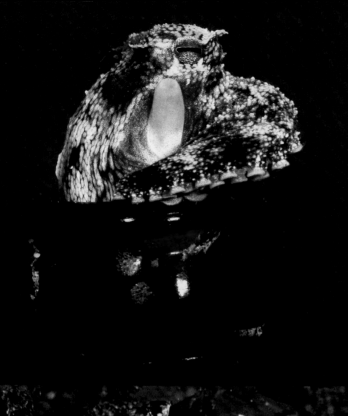

See-through security

As discussed in the previous section, the Veined octopus, now popularly known as the Coconut octopus, has acquired cephalopod fame for its antics in adapting the abandoned halves of the shells to use as portable shelters. As these pages illustrate, this enthusiasm for do it yourself (DIY) projects in home construction is not confined to coconuts. And, it would seem that octopuses in their material selection are quite comfortable in being fully transparent in their endeavours. This is most evident in the lower two photos where an octopus seems to favour cramming itself into the narrow neck of its chosen vessel. The photo opposite shows a Veined octopus addressing its talents to use the abandoned shell of the Grey bonnet gastropod together with a broken fizzy drink bottle. It looks a little optimistic that this combination will form a fully serviceable shelter from such disparate parts. However, having taken the effort to track down these attractive items of potential protection, it was clearly reluctant to release them for other seekers in this competitive market. The top left photo shows a juvenile octopus with more grandiose ambitions in home ownership adopting a whole beer bottle for accommodation though, in this case, the acquisition could hardly be described as portable.

Given the apparent vulnerability of its soft flesh, it is impressive that octopuses seem to have no problem dealing with the razor-sharp edges of broken glassware. This is evident in the right hand photo and those below. They bring to mind an incident with a dive knife. A diver, oddly intent on communicating with an octopus by waving his knife in front of it, did so by gingerly holding the blade fearful of inflicting a wound. The octopus promptly relieved him of his weapon by immediately grabbing the handle, creating the bizarre scene of a diver retreating from an octopus armed with a dive knife.

Location: Maumere, Flores Island, East Nusa Tenggara

Vanishing act

An encounter with a Bobtail squid, a tiny but exquisite jewel among the class of cephalopods, is one of the delights of diving at night. Bobtails typically measure no more than 3 centimetres, which makes them nicely bite-sized for a whole range of nocturnal prowlers. They are consequently very wary and adept at rapidly burying themselves in the sand, as shown in this sequence of photos. If the sand or sediment is sufficiently fine, they can literally vibrate their way into it. This vanishing act is completed with a final smoothing flourish by their upper pair of tentacles, which are then withdrawn below the surface too. If the substrate is of coarser material as shown on the right, it takes them a few more deft manoeuvres to place the final covering of gravel. The reverse process in their re-emergence is shown in a clockwise sequence starting from the right. Their favourite prey is small shrimps though the one sitting behind the squid top left may have found a temporary place safety.

These photos also illustrate the diverse range of colours and patterns that the Bobtail can produce. In an incredible example of symbiosis, these squid control the emissions of bioluminescent bacteria living within them called *Aliivibrio fischeri*. These enable the squid to create their own version of an invisibility cloak to completely camouflage themselves in moonlight.

Location: Kalabahi Bay, Alor, Lesser Sunda Islands

Dressed to the nines

How decorator crabs use adornments of living cnidarians is described in 'Changing the guard' in Chapter 3 and in 'Pirates and bonsai anemones' overleaf. These are compelling examples of what may be termed as aggressive decoration since it arms the subject with potent stinging cells. Those featured here are more complex since, while stinging cells are present, they are displayed as part of a whole wardrobe and illustrate the levels of extravagance to which some species of decorator crab will go. The main objective appears to be to obscure and confuse by creating an uncrab-like appearance rather than an overt statement of aggressive protection as boldly expressed by the individual on the red feather star on page 77. The wide range of decorations that these crabs use is all more incredible since they are composed of other living invertebrates. Given the efforts required to maintain all this ostentation, it could be wondered if they are trapped in an evolutionary path of increasing complexity. It demands high expenditure of time and energy as the crabs must first find and then attach this bewildering menagerie of adornment. And this investment does not come with a single payment. On every occasion that a crab moults, it emerges pristine from its old carapace and quite naked of decoration. It is, therefore, obliged to re-install its living array of attachments from scratch each time, and crabs moult up to 30 times during their lifetime. The decoration also has to be completed expeditiously since the new soft shell needs to harden and the crab is then at its most vulnerable. Presumably, though not widely reported, the optimum solution is to recycle its previous decoration by transferring it from the old shell. Such a process is illustrated in 'Changing the guard' though in that case, at least, the moulting crab remains surrounded by its decorative army of protective hydroids during the transfer.

In the theme of disruptive decoration, one of the more extreme examples is shown on the right, where the crab has used a rich cornucopia of ornamentation. These captive passengers include soft corals, zoanthids and another cnidarian called a coralmorpharian, which, although soft-bodied and anemone-like in appearance, is more closely related to hard corals. The soft corals and zoanthids are carefully arranged on its walking legs, leaving its front pair, with the claws, unencumbered. The crab has arranged the vividly coloured coralmorpharians on its carapace to create an impressive headdress. The individual top left has adopted a more ad-hoc approach appearing like some refugee from a raid on a vegetable stall. Along with a random sponge on the right, it has emphasised the eccentricity with three clumps of colonial ascidians, which look like bunches of broccoli. Sprigs of branching hydroids are favoured for the walking legs, and the whole ensemble is jauntily set off with a long feather of the same hydroid. Though in tune with this asymmetry, the crab below has kept it simple by fashioning just a sponge as a bizarre bonnet. This also has the benefit of being easy to re-size since all the crab needs to do is trim the sponge to fit, making it a quick-change artist in decorative moulting.

Location: Alor Island, East Nusa Tenggara and Lembeh Strait, Sulawesi

Pirates and bonsai anemones

Kleptoparasitism is a feeding strategy where food collected by one species is stolen by pirate species. Examples have been extensively documented and are widespread throughout the animal kingdom. It is practised by birds, notably by frigate or pirate birds, and a variety of insects such as ants which enslave captive species. The basic objective is the gaining of a free meal by the pirate. A possible marine example by the crinoid shrimp is considered in 'Clinging together' in Chapter 7. Combining kleptoparasitism with imposed diet control and growth limitation on a captive species is more unusual. A recent study has revealed this new development in which the victim is not only robbed but also regulated in size. Two examples of this practice by kleptoparasitic crabs are featured here. The so-called 'Teddy bear' crab is shown on the right and the more commonly encountered species, known as a Boxer or Pom-pom crab, is pictured top left. They both hold a pair of tiny captive sea anemones in their claws. These crabs essentially maintain the enforced symbionts in a 'bonsai' condition, which are conveniently carried around as tools to trap the crab's food. The anemones being armed with stinging cells can also help to immobilise prey making an easy meal for the crab while also providing it with a measure of protection. Indeed, the Boxer crab, when approached, will, instead of retreating, stand its ground and menacingly wave its pair of anemones in a fashion reminiscent of boxing gloves, hence its common name. When the anemone succeeds in acquiring food, the crab will dislodge it for its own consumption by using rapid leg movements. The study also found that when captive anemones were removed from the crabs and allowed to grow independently, they readily began to thrive, tripling in size as well as changing in morphology and colour.

As the crab in top left photo is an egg-laden female, this indicates that the anemones are captured on a long-term basis, which can extend through the brooding phase. It also underlines the inherent dexterity of these crustaceans since they are clearly able to keep their large egg mass safe from the stinging cells while maintaining the anemones in very close proximity.

Certain species of hermit crabs attach larger anemones to their adopted shells, as illustrated in the bottom left photo. This is a more mutualistic symbiotic relationship where both participants gain an advantage. The anemone's stinging cells provide the crab with extra protection and the anemones, which are quite omnivorous, gain high mobility giving them much wider access to food. Other species of crab use various cnidarians as decoration, examples of which are described in this chapter and Chapters 3 and 7.

Location: Pantar, South Banda Sea

CHAPTER 7

SYMBIOSIS

Constructive synergy

Symbiosis or 'living together' embraces an immense range of key relationships between species and on which the entire ecosystems of coral reefs depend. There are three main types—**mutualism** in which both parties benefit; **commensalism** where one benefits and the other is unaffected; and **parasitism** where the associate benefits at the expense of the host. These are not fixed categories, but there is more of a spectrum of benefit and loss across them. For coral reefs, symbiosis is most fundamentally evident in the mutualistic relationship between the corals and microscopic algae known as zooxanthellae. Various examples of the three main types of symbiosis are included in this chapter.

Small invertebrates, notably crustaceans, are very vulnerable to predation and many have evolved symbiotic relationships as survival strategies. This vulnerability is particularly relevant out in the open on the sandy slopes that adjoin coral reefs. Here, a common tactic is to build refuges in the sand. Pistol or snapping shrimp are skilled burrowers and, as can be seen in the photos, have robust sets of claws ideal for excavation. However, such an occupation comes with serious risks, especially for a shrimp with poor eyesight.

The relationship presented here features the mutualism between certain species of shrimp and goby fish. Burrow construction demands attention to detail and constant maintenance. But, such activity inevitably attracts attention from potential predators. The relationship provides the shrimp with very effective guardians since the gobies have excellent eyesight and are constantly vigilant. It may be that during the long process of evolution, the shrimp's vision, being less relied upon, has progressively deteriorated. The fish get the benefit of refuge in a burrow, which they would be unable to create themselves. There are usually male and female pairs of both fish and shrimp. The burrows include a breeding annex for the gobies and the energy of their mating rituals can cause extensive collapse, making it an extra busy time for the shrimps. So, the shrimps get all the heavy lifting work, but given the number of predators, the ceaseless surveillance required from the fish must also be demanding. To get close enough to observe this fascinating process in detail takes considerable patience and can easily take most of a dive. The gobies react to the slightest movement by immediately alerting the shrimp to retreat and often join them in the burrow. After eventually re-emerging, the gobies make a careful survey before signalling to the shrimp that it is safe to proceed with more excavation. The fish communicates through various fin movements. Flicks of the tail warn the shrimp to stay submerged or, if already out, will promptly send it below. As may also be seen in the photos, the shrimp are careful to maintain an additional line of communication by continually keeping an antenna in contact with the fish.

Location: Tulamben, Bali and Lembeh Strait, Sulawesi

Ménage à trois

The relationship between anemonefish and anemones is a classic example of symbiosis. Among other benefits such as cleaning services from the fish, each partner provides protection for the other. The anemonefish aggressively defend their host from predators such as butterflyfish. The association is very selective since, while there are many hundreds of species of anemones, only 10 are symbiotic with the 26 species of anemonefish. The fish are obligate partners meaning that they cannot effectively survive without the anemone. However, the relationship has been considered to be more facultative or optional for the anemones. Some potential evidence of this is discussed in 'Anemone city' in Chapter 1. Anemones are rapacious omnivores and, with their tentacles extensively armed with potent stinging cells, readily capture and consume fish. Despite this lethal threat, anemonefish have developed a process of acclimatisation, which enables them to swim freely among the otherwise deadly tentacles. The nature of this amazing symbiosis has been the subject of much debate and is still being researched. One suggestion is that anemonefish combine their inherently thick external mucous with that of the anemone to create an effective barrier against the stings. Current studies include assessment of the role of visual and chemical cues used for host recognition and the bio-chemical changes that occur during acclimatisation. This acquired immunity by the anemonefish needs to be regularly refreshed since, if isolated from its anemone, the protective relationship deteriorates and the fish risks getting fatally stung and consumed upon returning to its host. To avoid this, the fish adopts an elaborate behaviour, which involves an initial tentative touching of the tentacles—first with the fins and gradually extending over the body. This process can take several hours, but eventually the fish is able to fully immerse itself within the tentacles.

The intimate physical contact of anemonefish with anemones is well documented. However, the basis of the behaviour illustrated in the top left photo, where the fish has taken one of the tentacles well into its mouth, is unclear. One imaginative suggestion is that the fish could be trying to dislodge a tongue parasite, an example of which can be seen in the lower photo. In another remarkable relationship, these parasitic isopods enter fish as males and attach themselves to the gills. The male isopods are relatively small, but when established, one of them changes sex and then, as a functioning female, moves from the gills to attach itself to the fish's tongue. Here, it grows, gradually consuming and eventually replacing the tongue. Males then mate with the female in-situ and the resulting larvae subsequently emerge to seek a fresh host, and thus, perpetuate this bizarre life cycle.

Location: Komodo National Marine Park, Lesser Sunda Islands

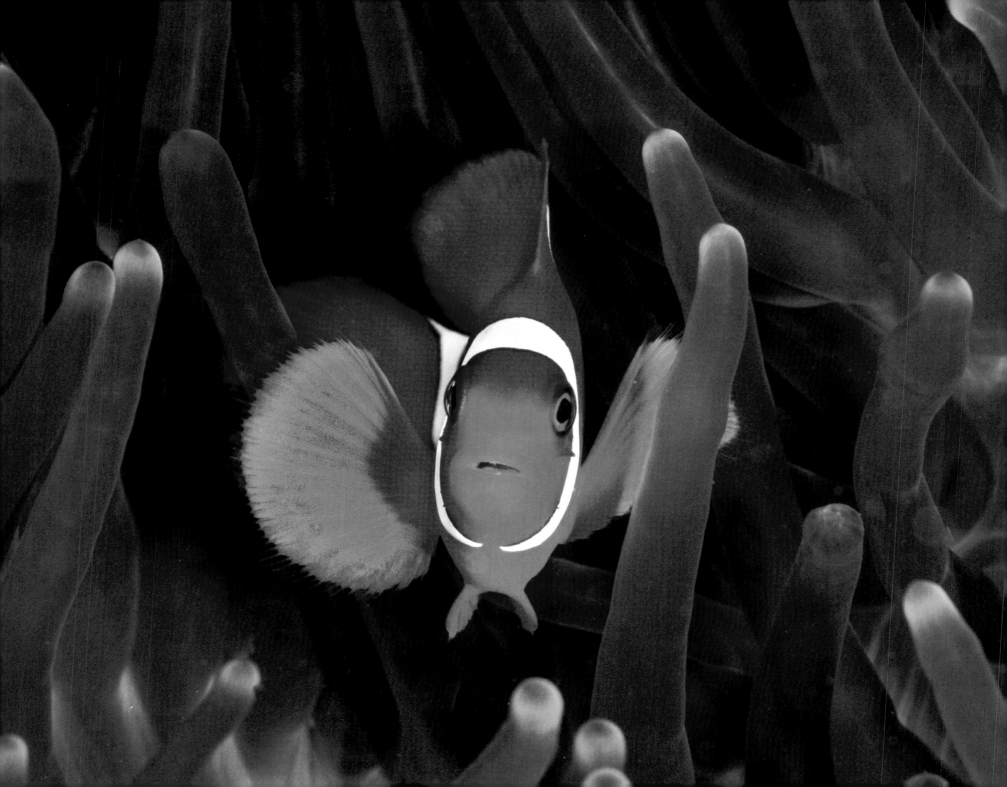

Gardening in a ring of fire

A male and female pair of Coleman shrimp, shown right, sit ringed by the venomous spines of a Fire sea urchin. The scene could be said to have geographical resonance in miniature since the Coral Triangle forms the eastern end of the region known as the 'Ring of Fire'. This stretches from Java around the northern Pacific rim to the Americas and is tectonically very active. As noted in 'Under the volcano' in Chapter 9, the Coral Triangle owes its very existence to this geological instability.

The distinctive shrimp are named after Neville Coleman, the irrepressibly enthusiastic marine naturalist, who discovered them in 1974 on the Great Barrier Reef. They associate exclusively with just one species of urchin and have since been found extensively within the Coral Triangle. I had the good fortune to meet Neville in Sipadan, Malaysia towards the end of his diving career in 2004. There, surrounded by the sharks, barracuda and turtles for which this location is famous, his focus was totally devoted to discovering new species of small invertebrates and especially his beloved nudibranchs. Coleman shrimp blend so well within their sanctuary that they are easily overlooked if not carefully inspected at close quarters—a proximity understandably avoided by some. I can personally confirm that the painful stings inflicted by the sea urchin's spines leave no doubt about their fiery nature. And it is easier to get stung than one might expect. The urchins can be very mobile and while concentrating on the urchin with the shrimps, one may be unaware of the approach by another. One's slightest movement can then result in an unforgettable engagement with the spines. The shrimp graze on parasites and algae and, while this housekeeping may be of some benefit to the urchin, these symbionts are also very dedicated gardeners. As can be seen in the photo, they have cleared a patch on the urchin's surface by locally snipping off the spines and tube feet. Ensconced within such potent protection, they can convey a strange sense of serenity—if you can avoid the spines.

Since these sea urchins are so visually prominent when around, virtually shouting aposematic warnings, one might expect that the shrimp should accordingly be easy to find. However, the presence of both is seasonal. Neville Coleman described these urchins as elusive and they are often completely absent in many locations. Even when they are present, the shrimp may not be in evidence. The investigation of many urchins may not result in even a single glimpse of a shrimp. I remember encountering hundreds of these urchins on a shallow plateau in the Banggai archipelago. A rare and quite amazing sight in itself, but an extensive search by the group of divers did not reveal even one Coleman shrimp. Once found, however, they make wonderful subjects. Not surprisingly, other small passengers such as the Zebra crab, pictured left, exploit the safe haven provided by these urchins and, as shown by the gravid female in the lower photo, are happy to adopt it as a convenient nursery.

Location: Ambon Harbour, Maluku

Candy crabs

This soft coral crab is the only member of the genus *Hoplophrys* and commonly known as the Candy crab from its likeness to that confectionary. Measuring less than 2 centimetres, it is a spectacular jewel of a crab even within a group that features some impressive visual confections, as shown in 'Changing the guard' in Chapter 3 and 'Dressed to the nines' in Chapter 6. It was first described in 1893, but perhaps surprisingly, given its undoubted attractions and unique status as the single species in the genus, there has been very little study since. As can be seen in the series of photos, the dendritic soft corals with which it exclusively associates have a range of colours, which are echoed by the crab. Its body form repeats the spiky nature of the polyps and its striped markings match the calcareous spicules in the coral branches.

As is also evident in the photos, the crab completes its disguise by decorating itself with the living polyps of the coral, making it almost impossible to detect unless it moves. An extra and inadvertent piece of decoration is present in the right hand photo. A sea urchin in its planktonic stage, with its long paddle-like appendage reaching upwards at 45 degrees, has attached itself to one of the crab's back legs. The female crab shown lower left is brooding a clutch of eggs. The symbiosis with its coral host is not well understood and could be either mutualistic or just commensal.

Location: Palau Mansuar, Raja Ampat, West Papua

Symbiotic embedments

The complex ecosystems of coral reefs harbour an immense diversity of species. Hard corals are known to host almost 900 organisms, more than a third of which are crustaceans. Some of these are parasitic symbionts and live their lives literally embedded in the coral. Many of them enter the coral during their settlement stage as larvae, and thus, must grow in tandem with their coral host though how they regulate the size of their enclosure throughout their life cycle is not fully understood. It may include the use of acid to dissolve the limestone skeleton of the coral and/or mechanical abrasion of the coral surface. A few examples of embedded crabs, bivalves and worms are included here. Some are amazingly colourful like the iridescent scallop shown opposite, which at around 5 centimetres stand out quite dramatically. Others are both tiny and cryptic and the Gall crab shown top left, often less than a tenth of the size of the scallop, is a prime example. While they may be quite common, knowledge of their associate coral species greatly aids their discovery. The hermit crab shown lower left having a fixed abode in the coral, is distinguished from the more familiar species which carry discarded snail shells around their backs and must seek larger ones as they grow. The bottom photo shows the delicate feathery spirals of a Christmas tree worm. Their plumes come in a wonderful range of colours and can be rapidly retracted into the coral with the entrance to their tube protected by the operculum cap shown in the lower centre of the photo.

Location: Kai Islands, South Banda Sea

Clinging together

Crinoids or feather stars provide a prolific and colourful presence within the coral reefs of the Coral Triangle. Though modest in comparison with corals, crinoids support a varied ensemble of symbionts, including fish and crustaceans, examples of which are featured here. Crinoids feed on plankton and are often seen perched high up on the reef, advantageously positioned in the currents. Some examples in full bloom are shown in 'The first link' and following pages, as they compete with anthias for plankton. Despite appearances, crinoids, like other species of sea stars, have five arms. However, being so extensively branched, they give the impression of having a far greater number. As filter feeders, their food is conveyed downwards along the arms. Consequently, their mouth as well as the anus is located on the upper surface of the body (known as the theca). This anatomical distinction departs from other echinoids in which the mouth is necessarily positioned on the underside, as they are grazers. Crinoids have an ancient lineage dating back some 480 million years to the Ordovician era. The widespread development of the flexible arms that we see today, occurred during the early Triassic period (about 230 million years ago), making them contemporary with modern corals. It is interesting to note that, in contrast to corals, they seem far less sensitive to global warming and appear to be still thriving in the face of climate change.

The crinoid clingfish shown right may be a male and female pair or perhaps, noting the confrontational pose, this could be the scene of a territorial dispute. These 3 centimetres long fish, which feed on tiny crustaceans, get their name from the thoracic disk, which they use to literally cling to the crinoid. It is quite a useful adaptation with a host that favours exposure to strong currents and fish, unlike crustaceans, do not have claws with which to hang on. The disk which is clearly visible in the right hand fish is formed by highly modified pelvic fins.

The most common symbiont associated with crinoids is the commensal shrimp shown top left. Crinoids have a wide variation in colour and pattern and the shrimps closely match those of their host. They are, thus, nicely camouflaged to wander along the feathery arms where they are well placed to intercept food particles snared by the host. Apart from such a tendency towards kleptoparasitism, the shrimp is also suspected of occasionally dining on the crinoid's tube feet, which would make the symbiosis decidedly parasitic. The Striped snapping shrimp, shown lower left, is usually found in mated pairs, which generally stay well concealed within the crinoid close to the theca. Another common associate, shown in 'Antennae range' in Chapter 3, is the Crinoid squat lobster. These also venture out along the arms and are camouflaged accordingly.

Location: Raja Ampat, West Papua and Lembeh Strait, Sulawesi

Taken to the cleaners?

The relationships between animals and those that clean them by removing their ectoparasites are some of the most well-known examples of symbiosis. Those between birds and mammals are prominent terrestrial ones, but by far, the most varied and widespread occur with marine fish. Early research conducted in the Bahamas in the 1960s emphasised the fundamental importance of these relationships. The health of fish populations and consequently, the very viability of the entire ecosystems of coral reefs were considered to depend on them. In the study, all of the cleaners on two isolated patch reefs were systematically removed. Within weeks, most of the fish had left the reefs and those that remained looked sadly unhealthy. The underlying message was widely and readily accepted. After all, anyone who has kept an aquarium, particularly with tropical marine species, will be aware of the critical need to control parasites. However, what happens in aquariums, whether in homes or laboratories, may depart quite significantly from the natural world. The key conclusions of the original work have been continually challenged in subsequent studies. The relationships were found to be a lot more complex and nuanced. One somewhat neglected factor is that the relationship is not dependent on just two players, but is triangular involving the parasites too, as noted earlier in this chapter in 'Ménage à trois'. In 1979, the marine biologist, G. S. Losey, speculated that some cleaners were nothing more than clever behavioural parasites taking advantage of the tactile stimulation enjoyed by client fish during the cleaning process. And this was not just a reference to false cleaners such as described in the 'Midas touch' in Chapter 2. It transpires that it is not only fang blennies that exploit the cleaning symbiosis by taking bites of healthy flesh. It may well be that some hitherto 'bona fide' cleaners are also taking their clients for a ride. The following decades have seen scientific opinion shift from the original hypothesis of an idealised mutual cooperation towards parasitism and the subject continues to be one of the ongoing and contested debates.

The subject is further explored in the next article 'Cleaning interrupted'. Beyond any controversy, the relationships that have evolved in cleaning are truly fascinating. Indeed, by maintaining a discreet presence, one may be treated to the combined pleasure of being able to approach reef fish attending cleaning stations much closer than is normally possible. The small striped wrasse are the most common and obligate cleaner fish, while butterflyfish and the juveniles of other species of wrasse attend to larger clients such as manta rays, as described in Chapter 2. Certain species of shrimp also provide cleaning services, as shown overleaf with the moray eel. The photos on this page and opposite were all taken at a well-established cleaning station managed by a pair of wrasse that concluded each session with an idiosyncratic flourish. The photo on the right captures this moment as they complete their services with a kind of 'high five' kiss. They did this with a variety of clients although with the large pufferfish top left, it was clearly a bit of a stretch. By way of variety in their performances, they sometimes finished with a lateral manoeuvre, as shown in the bottom left.

Location: Lembeh Strait, Sulawesi

Cleaning interrupted

Cleaning stations provide wonderful opportunities to observe behaviour in the interactions between marine fish. The clients seeking these services follow an elaborate range of protocols to communicate their desires to the cleaners. These signals include specific colour changes, extending jaws, flaring gill plates, and holding eccentric vertical positions, as displayed by the group of surgeonfish at the centre of the right hand photo. An example of differential colour shading is shown below by the unicornfish at a cleaning station. Here, a juvenile Redfin hogfish seen just in front of the leading fish is also in attendance to clean along with the more usual striped cleaner wrasse. The display of banded shading by the unicornfish may also help the cleaners to spot the ectoparasites.

However, as noted in the preceding article, 'Taken to the cleaners?', all may not be as it seems in these apparently idyllic scenes of cooperation between species. One clue comes from the trance-like state that the client fish display during cleaning, strongly indicating how much they enjoy the tactile stimulation. Such an addiction would make them vulnerable to exploitation by the cleaners which may remove more than just parasites. Another factor relates to the overwhelming ratio of clients to cleaners. The fusilier on the previous page, shown being cleaned by two wrasses, was from a huge shoal. It enjoyed a comprehensive service for over half an hour during which not one of its numerous companions got so much as cursory attention. The ratio of cleaners to fish and the time available are inadequate to service the majority of fish. This raises the question of whether such parasite control is really that critical to the health of fish populations overall. Moreover, the cleaning services with their established etiquettes are themselves quite vulnerable to disruption. As evident in the photos here, this may be caused by the arrival of large predators such as sharks or trevallies or, in the case of the scene on the right, by the dense shoals of fusiliers overwhelming the cleaning station and thus frustrating both the cleaners and the hopeful surgeonfish.

Location: Komodo National Marine Park, Lesser Sunda Islands

An odd couple

While gazing intently at a plume worm, my attention, sensing another presence, was diverted to a brown frogfish sitting unobtrusively in the background. Suddenly, in its curiously endearing but ungainly way, it began moving forward. As noted in 'Captivating mimicry' in Chapter 2, frogfish are poor swimmers and when they do move, it is typically by walking on their pectoral and pelvic fins. Here, the plume worms lay directly in its path. The feathery fans of these worms perform vital functions, acting both as lungs and for filter feeding. As they also present an attractive snack for many reef fish, the worm has evolved the defence of rapid retraction into the safety of its tube embedded in the substrate. So, with the looming proximity of the frogfish, I naturally expected to see the worm abruptly withdraw. Surprisingly, this did not happen—even when, as shown on the right, the frogfish, with a blatant lack of neighbourly etiquette, deliberately planted a pectoral 'foot' right on the extended plumes. This improbable scene unfolded with frogfish progressively nestling ever closer to the worm. Then, consolidating this bizarre process of inter-species bonding, the frogfish extended its lure. This incident raises some interesting questions:

- The advantages of the relationship to the frogfish are apparent: (a) For an ambush predator, sitting adjacent to a plume worm provides very effective camouflage. Since frogfish change their colour and markings to match their surroundings, it is feasible that its brown hue had been specifically adapted to harmonise with the plumes of the worm. (b) Since the plumes are attractive to the natural prey of the frogfish, this could tempt more into its own strike zone. Such prey would be less guarded than normal as they would hardly expect an attack to come from a plume worm. So, for the frogfish, this would appear to be an excellent arrangement.
- The advantages to the worm are more speculative. The frogfish and the worm do not compete for the same food. So, by tolerating the presence of the frogfish, the plume worm, at no cost to its diet, has acquired a guardian. Other potential advantages to the plume worm could be extended feeding periods plus energy conservation since it would, under frogfish protection, be able to significantly reduce the number of defensive withdrawals. While it is amazing that such a relationship would be credible, these photographs would suggest that there is tangible evidence in the sequence shown.

Location: Lembeh Strait, Sulawesi

A double mystery

These photos were taken during an encounter with four whale sharks. The group consisted of two males and two females. They were all juveniles attracted by the free meals of baitfish that the local fishermen hand out from their floating fishing platforms known as bagans. This opportunism by such juveniles has been well documented and is also described in 'The wonders of whale sharks' in Chapter 2. However, to encounter a group anywhere with an equal number of males and females is very unusual. The whale sharks in such groups have been found to be almost always all males, which is curious since the gender ratio at birth has been recorded as roughly equal. As confirmed by the absence of claspers at the ends of the pelvic fins in the lower left photo, the individual featured here is of one of the two females. As is also very evident, this fish had the extra distinction of being adorned by over 80 remoras. While small numbers of remoras are commonly seen as commensals on large pelagic fish, a huge aggregation like this is rather extraordinary and presents a second mystery. The other three whale sharks had but half a dozen remoras between them. The female with just a few remoras generally kept below 20 metres and was not observed to visit the bagan and was never seen near the surface. These different behaviours raise many questions.

- Why are sightings of young females so rare?
- Do they generally stay deep and beyond easy observation and diving range?
- What caused the huge aggregation of remoras to concentrate on just one of the whale sharks?
- Was such behaviour perhaps related to a spawning event of these commensals?
- What made the female with all the remoras act with such apparent confidence, in marked contrast with the other female, and join the two males in feeding at the bagan?
- Did so many remoras influence its behaviour, perhaps making it feel more secure enveloped within a large if rather strange shroud?
- Alternatively, could it be that the presence and activities of so many remoras had adversely affected this whale shark's normal feeding thus making it unusually hungry and therefore far more inclined to succumb to the temptation of the free meals on offer at the bagan?

Location: Triton Bay, West Papua

CHAPTER 8

REPTILES

A breath of fresh air

The Golden sea snake seemed briefly held in suspended animation as it floated across an azure blue sky in the crystal-clear waters of the Banda Sea. This species can grow to well over a metre in length and is reputed to have the most potent venoms of all snakes—marine and terrestrial. It swims using its paddle-like tail and, while well adapted to an aquatic life, does not have gills and so, must return to the surface periodically to breathe. It was this necessity that is captured in the image on the right. However, such photogenic opportunities require a little luck since, in their feeding activity, sea snakes can reach diving depths between 50 and 100 metres and remain submerged for more than 2 hours. When they come to the surface to breathe, it is only for 30 seconds or so. Photoreceptors in the skin of the sea snake's tail enable it to sense light intensity. It has been proposed that this enhances its ability to hide within rock crevices or coral during the day to avoid predation by sharks. Despite such protective biology and its potent reputation, it remains a species under threat and is on the IUCN Red List. Nevertheless, the individuals featured here appeared to have no immediate concerns about survival and were moving freely about the reef during a beautifully bright sunny day.

The individual pictured on the right, after its momentary visit to the surface, joined other companions searching around a large sea anemone for prey such as small fish and crustaceans. The lower left photo shows it leaving the scene and the no doubt relieved anemonefish returning to its host anemone. The location features the volcano of Gunung Api, and is renowned for the unusual abundance of sea snakes that it supports. So, the anemonefish would be accustomed to such predatory investigations. Certainly, during my observations, it patiently endured the attention of four different snakes. The anemonefish was possibly more concerned about the condition of its anemone which, as shown, was suffering from extensive bleaching although there was no bleaching evident in the adjacent corals. Like Batu Tara featured in Chapter 9, Gunung Api is an active volcano and has also been reported to have rich and rapid coral growth around the lava flows.

Location: Gunung Api volcano, Banda Sea

Resident evil

The Komodo dragon is the world's largest lizard weighing up to 70 kilograms and attains lengths exceeding three metres. They are now restricted to a small area in eastern Indonesia, which includes the Komodo National Park. Their isolation makes their range the most restricted of any large apex predator. Though not aquatic, they are strong swimmers and have been encountered miles out to sea travelling between islands. They were quite unknown in Europe until 1910. That such seemingly 'prehistoric' beasts were still living was exciting news. Even in the 1950s, they were still associated more with myth than reality until one was featured as the first star of David Attenborough's famous ground-breaking BBC television programme series, Zoo Quest. These fearsome reptiles enjoy an evil reputation, which includes human prey, grave robbing and very unsavory killing methods. Their saliva was thought to contain masses of bacteria condemning bitten prey to a lingering death. Research, however, has now established that they actually have a potent venom in glands in their lower jaws and are considered to have evolved from a group comprising the largest ever venomous animal—the 5.5 metre long, 600 kilograms *Megalania* lizard. The venom is a complex cocktail of toxins, which causes blood loss, tissue damage and rapidly lowers blood pressure rendering victims into shock. So, once bitten, prey as large as water buffalo, which may weigh more than ten times that of a dragon, can be tracked at leisure until they are too weak to defend themselves.

While their malevolent notoriety is perhaps overstated (for instance, only four human deaths have been recorded in the last 40 years), science continues to reveal surprising sophistication in their biology. For example, in a process known as parthenogenesis, which is rare in vertebrates, female Komodo dragons are able to breed without male insemination. Despite their 'slow-death' tactical ability, noted above, Komodo dragons mainly consume carrion. Their long, forked tongues collect airborne molecules, which are then analysed by the Jacobson's organ located in the roof of their mouth. This organ is so sensitive that it enables rotting flesh to not only be detected from miles away but also from which direction the scent emanates. It was this principal diet that led to the erroneous belief about the toxic bacteria, but it was also a puzzle why the dragons did not get ill. It now emerges that their blood is loaded with anti-microbial peptides (AMPs), which provide a comprehensive immune defence. Research is now focussed on whether these AMPs can be developed to combat antibiotic-resistant bacteria in humans.

Location: Horseshoe Bay, South Rinca, Flores

Turning turtle

The two species of marine turtle most commonly encountered in the Coral Triangle are the Hawksbill and the Green turtle (which is featured in the following section). They are both global in their range across tropical and sub-tropical regions. The females of both species, having an incredible navigational ability, return to lay their eggs in the same breeding grounds from which they originated. As noted overleaf, this creates both threats and opportunities for the survival of these wonderful but endangered animals. The array of threats faced by both species include the loss of feeding and nesting habitats, commercial harvesting of the eggs, and as bycatch in fishing nets and lines. As highlighted by conservation organisations like the World Wildlife Fund (WWF) and Conservation International (CI), one of the greatest threats still comes from the illegal trade in wildlife. Despite the implication in the name, the primary source of the tortoiseshell widely used for decoration is not obtained from the eponymous land-based reptiles, but from sea turtle shells. This applies particularly to Hawksbills even though the international trade in their shells was prohibited nearly 30 years ago. The combined impact from these multiple threats has been harder on the Hawksbills, which are now listed as critically endangered by the IUCN. They are now protected by a range of international conventions such as that on Migratory Species (CMS) or for the International Trade in endangered species of Wildlife Fauna and Flora (CITES). Such international cooperation is essential for turtles given their extensive geographic range.

Marine turtles are the descendants of a group of reptiles that has inhabited the oceans for over 100 million years. As a key migratory species, they provide an important link in marine ecosystems and help to sustain the health and diversity of coral reefs. Hawksbills generally reach one metre as adults and weigh up 80 kilograms. Unlike the Green turtle, they remain omnivores throughout their life and are named after their distinctive pointed beak. They are particularly partial to sponges and will assume possession of an outcrop by sitting right over it and dining at their leisure, as shown in the lower left photo. They also feed on anemones and jellyfish and, on more than one occasion, I have needed to fend off a determined Hawksbill, which had mistaken the dome port of my camera housing for a potential meal. While presenting dramatic photographic opportunities, the likely abrasion inflicted by their sharp beaks would not have improved the optics of my camera system. Hawksbills also consume algae, sea urchins, crustaceans, and molluscs and are often seen in a duck-like orientation with their tails up and heads buried in the coral seeking such tasty fare, as shown in the top left photo. The scene shown opposite occurred towards the end of a dive. The turtle glanced at my companions high up on the reef and turned back to peer at me, perhaps wondering how I managed to remain with such a silent presence and making no bubbles—thus providing more of the joys of using a rebreather.

Location: Komodo National Marine Park, Lesser Sunda Islands

Becoming vegan

The Green turtle gets its name not from how it looks, but what it eats. As immediately obvious from the photos, they are not actually green—at least on the outside. Their name relates to the colour of their fat and cartilage caused by their diet of seagrass and algae. As adults, they are the only herbivorous marine turtle though they do not start that way. As part of their survival strategy, the hatchlings from all the nests at a site emerge simultaneously. Those that manage to survive the gauntlet of waiting predators in their desperate rush to the sea, disappear for several years in a pelagic existence on the high seas. During this period, in which they grow to around 25 centimetres, they are quite omnivorous, consuming crustaceans, worms and aquatic insects as well as plants. They eventually reappear in coastal regions as young juveniles. There they make a major lifestyle change, turning vegan and remaining inshore over the 25 or so years before reaching maturity. In the latter periods of their life cycle, the mature adults migrate back to the very breeding grounds from which they hatched. Green turtles nest in more than 80 countries and commercial hunters and poachers know when and where to expect their arrival. Fortunately, this knowledge also facilitates targeted conservational efforts to protect the turtles and their nesting sites. With the natural survival rate at only 1 in 1000, they certainly need our help. Only the females leave the water to nest in the sand above high tide level and can lay around 100 eggs every two to three years. One clutch of eggs may have several paternities since the females can store the sperm from multiple couplings in their oviducts.

Green turtles are one of the largest of the marine species reaching lengths of 1.5 metres and weighing over 250 kilograms. They play a key role in seagrass ecosystems since what they consume is recycled as nutrients for many other animals and plants. The seagrass beds also provide nurseries for many species of fish and invertebrates. The adults do not always stick to their vegan diet and are partial to the occasional jellyfish. Tragically, this also leads them to consume plastic bags. Since the mass production of plastic started in the 1940s, the effect on turtles and other large marine species has been devastating. So, it is vital to reduce all forms of this pollution which has reached epidemic proportions. Green turtles are listed as endangered in the IUCN Red List with a trend of decreasing populations.

The individuals pictured here were encountered during a single dive in the Komodo National Marine Park. They were quite relaxed and not wary of approach, which encouragingly indicated that the protection measures imposed for the park were being effective. Towards the end of the dive, as I travelled back along the reef, a Green turtle emerged dappled in the light of the low sun. As I moved towards it to capture a parting shot, a manta ray swam into view completing this magical vista of the coral reef panorama.

Location: Komodo National Marine Park, Lesser Sunda Islands

CHAPTER 9

CONSERVATION

Photo courtesy of Bruno Hopff, Amira

Climate change

While the prime focus of this book is to highlight and celebrate the amazingly rich marine biodiversity of the Coral Triangle, the urgent need for conservation is also a constant theme. The leading factor is the upward trend in global warming causing sustained peaks in sea temperature rise. The resulting thermal stress has led to devastating coral loss through mass bleaching as tragically reported in vast areas of the Great Barrier Reef (GBR) and the Chagos Archipelago in the Indian Ocean. The latter, having the protection of a military base, was previously considered to be one of the healthiest and most resilient examples of tropical coral reefs, yet it too has now succumbed to the ravages of climate change. When the full syndrome of stressors is included such as pollution, ocean acidification, sea level rise, habitat loss, and commercial exploitation, the prospects look decidedly grim. As noted in the first chapter, tropical coral reefs are some of the most biologically diverse ecosystems on the planet. In terms of the survival of coral reefs, this diversity is double-edged. It is crucial to their resilience, yet conversely, the very complexity of these inter-connected ecosystems makes any of our efforts to manage them protectively likely to fail. Our actions are prone to have unintended consequences. What we have witnessed during the two decades following the inaugural International Year of the Reef in 1997, is an increasing deterioration of coral reefs on a global scale.

So is there any hope? The top photo on the left shows bleaching of a hard coral with a transition from partially bleached polyps to those completely white. These have lost their entire complement of the symbiotic algae, which impart the coral's colour. In this case, it is hard to say whether the coral is deteriorating or actually recovering. Recovery is possible if the peak in temperature is of short duration and the coral can re-absorb its lost algae. The polyps of the fully healthy coral are shown in the close-up below the bleached example. The scene illustrated on the right conveys the stark reality of coral bleaching. However, although it looks like a coral bone yard, in many ways, it is not as bleak as it might initially seem. It is not mass bleaching and many parts of the surrounding reef appear healthy. It is also interesting to note that in this photo, the coral is bleached while the anemone in the foreground is not. In the bottom left photo, the reverse scenario applies. Corals and anemones are both cnidaria and the differential effects shown here hint at the underlying complexity of these inter-relationships. The right hand photo was taken towards the end of a four-week field trip during March and April 2016, travelling some 1400 kilometres eastwards from Sulawesi to Sorong in West Papua. En route, hardly any coral bleaching was observed until reaching Kaimana—perhaps significantly, the part of the trip closest to Australia where the GBR was then suffering major mass bleaching events. In Kaimana, both hard and soft corals were affected, but only in isolated patches. Upon returning to the same area a year later, it was heartening to see that the corals had substantially recovered despite still elevated sea temperatures of 30°C. To view these local observations as indicative reasons for a wider optimism is just speculation. Nevertheless, throughout my diving in the Coral Triangle, over the last decade, I have only encountered a few isolated cases of bleaching. So, as noted in the 'The deep diversity of the shallow reef' in Chapter 1, there are still grounds for hope.

Location: Pulau Molu, Southeast Banda Sea

Destructive fishing methods

All of the fishing methods discussed here are very destructive, especially when used in combination. It is important to recognise that although the practice is widespread, climate change presents a far greater and existential threat on a global scale. Destructive fishing is not just a local activity but also, and more insidiously, is compounded by outsiders. Ironically, the better local communities care for their reefs, the more attractive they are to poachers. The photos show the grim reality of devastated reefs—on the right by blasting and top left by cyanide fishing where the dead and dying coral is being overgrown and encrusted by algae. These tragic scenes are all the more poignant when compared to healthy reefs such as those of Kusu Ridge featured in Chapter 1. Even the living but weakened coral around the margins of the blast zones is vulnerable. For example, as shown in the bottom left photo, predation by the infamous crown-of-thorns starfish becomes more critical when the coral is stressed.

Blast fishing

Although blast fishing is now illegal throughout Indonesia, it is still very prevalent. The deep ominous booms can be felt underwater from a distance of over 20 kilometres. The homemade bombs use chemical fertiliser packed in glass bottles. The resulting pressure waves kill and maim fish within a blast zone of 20 square metres and are brutally effective on flat reefs down to a depth of 15 metres. The fish are collected either from the surface or by diving, sometimes using a crude hookah air compressor and a long hose, although this is more commonly used for cyanide fishing. Recovery from such devastation often takes decades.

Cyanide fishing

This is another illegal method and may be used as a prelude to blast fishing. Potassium cyanide is used to stun fish and other creatures, which are then easily collected. It can be disturbingly effective over a wide area in a relatively short time. Typical targets are wrasse, groupers, lobsters and aquarium fish. The poison has long-term residual effects slowing killing the reef itself and its inhabitants.

Reef-gleaning

This is probably the most widespread destructive fishing technique in the Coral Triangle. Metal rods are used as levers to overturn coral stands and rocks, breaking up the reef to enable collection of edible delicacies such as octopuses, crabs, shrimps, snails and sea cucumbers. Whole families, including young children, may be observed at low tide combing the top of a reef. This ongoing attrition destroys vital shallow habitats, which provide nurseries for a diverse range of species of both fish and invertebrates.

Location: Pulau Suangi, between the Seram and Banda Seas, Maluku

Unsafe harbour

In his seminal book of 1859, 'The Malay Archipelago', Alfred Russell Wallace recorded his delight upon seeing the marine life of Ambon's natural harbour. His lyrical description portrays a radically different picture than the sad scenes encountered there today. He wrote:

> *"Passing the harbour, in appearance like a fine river, the clearness of the water afforded me one of the most astonishing and beautiful sights I have ever beheld. The bottom was absolutely hidden by a continuous series of corals, sponges, actiniae, and other marine productions of magnificent dimensions, varied forms, and brilliant colours. The depth varied from about twenty to fifty feet, and the bottom was very uneven, rocks and chasms and little hills and valleys, offering a variety of stations for the growth of these animal forests. In and out among them, moved numbers of blue and red and yellow fishes, spotted and banded and striped in the most striking manner, while great orange or rosy transparent medusae floated along near the surface. It was a sight to gaze at for hours, and no description can do justice to its surpassing beauty and interest. For once the reality exceeded the most glowing accounts I had ever read of the wonders of the coral sea. There is perhaps no spot in the world richer in marine productions, corals, shells, and fishes, than the harbour of Amboyna."*

This account struck a poignant resonance when I first visited Ambon in 2012. The photos reflect the pervading sense of desolation. A pair of Ribbon moray eels portray the struggle for life as they strain upwards through the thick carpet of inert sediment, which spreads everywhere. The wondrous vision that Wallace beheld has long vanished as the harbour has been progressively overloaded with pollution. A cuttlefish hunts among mounds of rubbish where once used to be a rich coral reef. Plastic is an increasing and deadly menace to the marine environment and the image of a nudibranch cruising over an abandoned child's toy conveys a bitterly ironic message. The shocking pink colour, challenging the colourful livery of the nudibranch, and demonic twinkle in the duck's eye bring an almost theatrical impact to the inherent menace. The once rich and vibrant coral reefs are now replaced with vast stretches of featureless plains of deep silt. Yet, though Ambon harbour is now a mere shadow of its former glory, a surprising variety marine life still survives. This indicates a potential that it could be nurtured back to some form of recovery. Unfortunately, this seems currently unlikely with the ongoing destructive pressure that continues afflict this fragile environment.

Location: Ambon Island, Maluku

Exotic invaders

It is well known that biological invasions can critically degrade ecosystems and lead to loss of biodiversity. Once invasive species become established, their deleterious effects can spread unchecked throughout a region. Two localised examples within the Coral Triangle are discussed in 'Anemone invasion' and 'Endangered yet invasive' in Chapter 3. The first is a natural one, but shows how locally dominant the invader can become and indicates the potential for an unnatural escalation if the ecological balance is upset. This is more likely to occur when the initial introduction occurs unnaturally. This is illustrated with the Banggai cardinalfish in the second example. Although native within the Coral Triangle, they are naturally restricted to a small area in the Banggai archipelago. They are more of a prey than a predatory fish, yet where they have been artificially introduced to other areas, they have been seen to rapidly multiply in unnaturally high numbers. It is too early to say what the long-term effects of this may be, but it pales in comparison to the introduction of an invader from the Indo-Pacific region into the Caribbean, Western Atlantic and Mediterranean, which has rapidly become a scourge.

The two species of Indo-Pacific lionfish shown here, *Pterois volitans* (right) and *P. miles* (left), are commonly seen in the Coral Triangle. Although very efficient predators of smaller fish, they live there in balance with the overall ecosystem. However, beyond their natural environment, they are transformed into exotic aliens with many of the hallmarks of the ultimate predatory invader. Being unknown to the native fish has enabled them to become supremely voracious. They also prey on invertebrates, including high quantities of crustaceans. Lionfish have a cryptic appearance and are armed with defensive venomous spines. They have high rates of reproduction aided by early maturation combined with rapid growth. When this potent threat is added to the existing environmental stressors from pollution and climate change, it effectively amounts to a perfect storm of an invasion. Both species of lionfish are believed to have been introduced into the coastal waters of Florida and their presence there was first reported around 1985. Since then, their geographical spread has been rapid, extending across the Western Atlantic, into the Gulf of Mexico, up to Maine and down as far as Brazil. The effects on the native fish have been catastrophic with estimates on some reefs suggesting a loss of 95%. By 2004, they had become established in the Bahamas and during the next decade, were estimated to have colonised over 7 million square kilometres of the region. By 2014, the lionfish invasion had extended throughout the Caribbean islands with densities reported up to 390 per hectare and with their presence recorded to depths of 300 metres. One particularly alarming statistic indicated from genetic testing is that the invasion may have originated from less than half a dozen fish. It is the worst ever tropical marine invasion on record and there are no easy solutions. It has become so entrenched that eradication appears impractical and only some form of control is feasible. Only time will tell how the local ecosystems will respond to this onslaught.

The small majority and Darwin's Paradox

Shortly after his return from his famous voyage on HMS Beagle during 1831–36, Charles Darwin published his theory on the formation of coral reefs and, in particular, the origin of atolls. This was a mystery at the time and the subject was seen as a key aspect of the scientific investigations planned during the voyage. Darwin's brilliant breakthrough concept was about linking geology with marine biology. He postulated that the slow rise and subsequent subsidence of volcanic islands over millennia led initially to the formation of fringing reefs, eventually leaving an atoll as the remaining structure at the surface while the original now inactive volcano returned to the sea bed. It could be viewed as Darwin's first theory of evolution. Though significantly less contentious than his subsequent one on the origin of the species, it also remained a topic of hot debate until well into the 20th century. Darwin's theory was finally vindicated in 1953 when deep drilling into an atoll in the Pacific Ocean established that the coral structure did, in fact, sit directly upon volcanic rock.

Observations are most interesting when they cannot easily be explained. With regard to coral reefs, there is still one conundrum that Darwin's observations, made nearly two centuries ago, have left us to ponder. Known as Darwin's Paradox, it arose from the analogy of viewing coral reefs as fertile oases in the deserts of the ocean. How could it be that the incredibly rich ecosystems of tropical coral reefs could flourish in the crystal-clear but nutrient-poor waters that surround them? Of course, Darwin was not aware of the vital photosynthetic symbiosis between the microscopic algae and their coral hosts, which inherently relies on clear water, allowing the sunlight to penetrate. Even so, the whole nature of this relationship is still not fully understood, especially with the heightened interest in coral resilience in the face of climate change. But, photosynthesis is only part of the answer since the teeming diversity of life across the entire ecosystem needs explanation. Two recent hypotheses have been advanced to address this. One is known as the 'Island Mass Effect' in which the physical form of the reef develops to control the flow and circulation of the seawater. It, thus, enhances and contains the production of phytoplankton within the reef itself. A more recent proposal focusses on the often overlooked but extensive community of fish that could be termed as the small majority. While the many species of large and colourful fish such as wrasse, parrotfish and angelfish are typically associated with coral reefs, vast numbers of tiny fish contribute to up to 40% of the overall diversity of reef fish. Cryptobenthic fish, like the pair of blennies shown in the lower left photo, are key representatives of this biota. Other examples of the delightful small citizens of the reef include the Bearded goby opposite along with the green Halimeda ghost pipefish and the Purple dartfish on the left. Small fish tend to have high metabolisms and reproduction rates, and their larvae are not widely dispersed, but settle locally. It has been estimated that the cryptobenthic fish could produce around 60% of the fish biomass consumed on coral reefs. The small majority may, thus, make a big contribution to solving Darwin's Paradox.

Beneath the volcano

As described in Chapter 1, volcanos played a key role in the creation of the Coral Triangle. It is fitting, therefore, that this book should conclude with one. Batu Tara is an active volcano on the island of Komba located about 50 kilometres north of Lembata Island in the Flores Sea. With its summit rising around 750 metres above sea level, it is one of the smaller volcanoes in Indonesia—at least above the surface. Its base reaches down to the sea floor 3000 metres below. Nevertheless, it is an intimidating experience up close, especially when it is actively producing explosions of molten lava and belching plumes of ash from its crater about 900 metres in diameter. Now sitting in splendid isolation, it is the last of a group of 30 volcanos which formed one of the many such arcs that originally gave rise to the Coral Triangle. Indonesia is the world's most tectonically dynamic country and has over a third of its active volcanos.

Diving in the almost primeval environment at Komba provides a surreal experience in which one feels the very presence of creation. The tableau of the volcano and the coral reef conveys an image of how the Coral Triangle continues to be formed. One's awareness is further heightened should the volcano emit even a minor rumble during a dive. It is a physical sensation—felt rather than heard. As can be seen in the right hand photo, a vast landslide of volcanic material continuously flows down from the crater into the sea. This is dramatically evident below the surface as illustrated by the photos on the left. Despite the enveloping streams of ash that continually descend into the sea, life can be seen securing footholds in this nutrient-rich environment. These scenes echo the very conditions that originally led to the birth of the Coral Triangle and which, in turn, led to the development of its uniquely rich biodiversity.

What is perhaps surprising, given how sensitive corals can be to environmental change, is how quickly the marine life becomes established in such dynamically changing conditions. The top photo on the left was taken in the centre of the fan of ash. Here, a stand of *Acropora* coral has been toppled and partially smothered. However, a good proportion is still alive and continues to grow along with the other species of corals, which are forming an expanding buttress to the flow of ash. The bottom photo shows the increasing growth of corals along the margins of the ash flow. While some corals are smothered and die, new colonies continue to appear. Pink sea fans are finding purchase, anchoring into the hard corals. As one travels beyond the fan of ash, the proliferation of corals and other marine life progressively increases until one is surrounded by examples of those rich and varied coral gardens discussed in Chapter 1.

Location: Komba, Lembata Island, Flores Sea

Appendix

Photo courtesy of Bruno Hopff

Abbreviations

cf: Compare to, e.g. in this book where a definite species or genus identification cannot be determined from the photograph, but compares with one similar.
CI: Conservation International.
CITES: The Convention on International Trade in Endangered Species of Wild Fauna and Flora.
cm: centimetre(s)
CMS: Convention on the Conservation of Migratory Species of Wild Animals.
DNA (deoxyribonucleic acid): The molecule of heredity composed of nucleotides. It is the carrier of genetic information within a cell that is transmitted from generation to generation.
IUCN: International Union for Conservation of Nature.
mm: millimetre(s)
mya: Million years ago.
NOAA: National Oceanic and Atmospheric Administration.
Scuba: Self-contained underwater breathing apparatus.
TL: Total length of a fish. Straight-line measurement from snout tip to base of caudal fin.
WWF: World Wide Fund for Nature.

Equipment

Diving: Rebreathers: Poseidon MarkIV and Mark7. These are fully closed-circuit, fully automated systems.
Photographic: Cameras: Nikon DSLRs D7000 and D500.
Lenses: Macro: Nikon AF-S 105 mm f2.8G VR Micro, Nikon 60 mm F/2.8D Micro, supplemented with Subsee Underwater close-up lenses. **Wide angle:** Tokina AT-X DX Fisheye, 10-17 F3.5-4.5, zoom.
Underwater housings by Nauticam. Strobes: Inon Z240 and S2000.

Glossary

Actiniae: Sea anemone of the genus *Actinia*.
Allies: Organisms sharing an evolutionary relationship to each other.
Amphipods: Invertebrates belonging to an **order** of **crustaceans** having no **carapace**.
Antennae: Anterior appendages or 'feelers' of an **arthropod.**
Anthropogenic: Arising from human impact on the environment.
Anthozoa: A **class** of marine **invertebrates** which includes the sea anemones, **hard** or **stony corals** and **soft corals**.
Aposematism: Overt warning by an animal to potential predators that it is dangerous or unwise to attack, e.g. being venomous, toxic or foul-tasting. Often signalled by aposematic colouring.
Arthropoda: A **phylum** of **invertebrates** with an **exoskeleton**, a segmented body, and paired jointed appendages, e.g. a **crustacean** or spider.
Ascidian: A sac-like marine filter feeding invertebrate in the subphylum of Tunicata. Also known as **tunicates** and sea squirts.
Asexual: Reproduction by means other than sexual congress such as budding, cloning or **fission** (aka fissiparity).
Associate: Where typically two species associate in a **symbiotic** relationship. May involve more than two species as in cleaning relationships.
Bagan: Large wood-frame floating platform used for fishing in sheltered bays in Indonesia.
Benthic: Organisms which live at the bottom of a body of water which may move over, burrow or be attached to a substrate.
Biogeography: The study of the distribution of **species** geographically and through geological time.
Bioluminescence: The emission of light from living organisms.
Biota: The plant and animal life of a region or geological period.
Biramous: dividing to form two branches (particularly of crustacean limbs and antennae).
Brood pouch: An enclosure in the body of an animal where the eggs or embryos are deposited and undergo development until hatching.
Bryzoan: A **phylum** of aquatic colonial filter feeding invertebrates, typically about 0.5 millimetre in size.
Calcareous: Composed of calcium carbonate, e.g. chalky.
Calcification: Hardening of body tissue or other matter through the accumulation of calcium salts.
Caldera: A large depression, typically circular, formed when a volcano erupts and collapses.
Caprellid: Member of the **family** of **amphipods** commonly known as skeleton shrimps.
Carapace: Chitinous and/or calcareous dorsal (upper) section of an **exoskeleton** or shell.
Cartilaginous: Composed entirely of cartilage, e.g. skeletons of sharks and rays.
Cephalopod: Member of the **class** Cephalopoda of the **phylum** Mollusca, e.g. cuttlefish, squid, and octopus.
Chitin: Horny organic compound forming part of skin or shell, e.g. in some invertebrates and fish.
Chordate: An animal of the **phylum** Chordata which includes vertebrates such as fishes, amphibians, reptiles, birds, mammals and also **ascidians.**
Chromatic colour: A colour where a particular wavelength predominates.
Chromatophore: Pigment bearing cell in skin of, e.g. fish and cephalopods. The pigment may be selectively expanded or contracted by radial muscles, thus affecting colour variations. The overall colour effect is varied with the nervous state of the organism, e.g. during aggression and sexual arousal.
Claspers: Claspers are an external pair of appendages found on male sharks and rays which are similar to a penis and used during mating.
Class: See **taxonomic rank.**
Cleaning station: An established location on a reef where cleaners (either certain **species** small fish or shrimps) remove **ectoparasites** from visiting reef fish. The station itself may be an anemone, sponge or coral head or simply a ledge or crevice, which provide a refuge for the cleaners.
Cnidaria: **Phylum** of animals that includes corals, sea anemones and jellyfish. They have a radial symmetrical in two major body forms—the **polyp** and the medusa.
Coral bleaching: Discolouration of corals through loss of their symbiotic microalgae known as **zooxanthellae**.
Coral polyp: Soft-bodied attached stage of **Cnidaria**. Polyps have a central mouth or cavity surrounded by tentacles. They are one of the two forms in the **phylum Cnidaria**, the other form being the medusa. They are generally **colonial** which build structures to support themselves.
Colonial: Communal animals of the same **species** living together sometimes

physically attached as in **ascidians** or otherwise with social links such as bees.
Commensalism: See **symbiosis.**
Copepod: Member of crustacean subclass Copepoda. Copepods, being one of the most abundant forms of life on Earth, are very important ecologically.
Crinoid: Also known as **feather stars**, crinoids are marine animals of the **class** Crinoidea, in the **phylum** Echinodermata, which also includes sea urchins, sea stars and brittle stars.
Crustacean: **Class** of **invertebrates** with hard segmented outer shells such as crabs, lobsters and shrimps.
Cryptic: Ability of an animal to blend in its environment.
Cryptobenthic: Describing organisms which are both **cryptic** and **benthic.**
Cypraeidae: Family of sea snails, the common name for which is cowries.
Demospongiae: The most diverse **class** in the **phylum** Porifera. They include over 75% of all sponges.
Dendritic: Having a tree-like branching form.
Depth of field: In photography, the distance between the nearest and the furthest objects giving a reasonably sharp focussed image.
Diatom: Major group of microalgae in the **phylum** of Ochrophyta.
Echinoid: Echinoids or sea urchins belong to the **phylum** Echinodermata, which includes sea stars, brittle stars, sea cucumbers, and crinoids. Their hard shell is covered with small knobs (**tubercles**) to which the spines are attached.
Ectoparasite: A parasite that lives on the outer surface of an animal.
El Niño: A periodic climate event causing the warming of sea surface temperature. It happens every few years, and is typically concentrated in the central-east equatorial Pacific.
Endemic: Native and restricted to a defined locality.
Esca: The lure of a frogfish. Its shape, which may mimic a small fish or crustacean, is one of the main distinguishing features of a species.
Eukaryotes: Organisms whose cells have a nucleus enclosed within a membrane.
Exoskeleton: Rigid or articulated exterior covering of many animals, particularly **invertebrates.**
Facultative: A symbiotic relationship where the species live together by choice—cf. **obligate.**
Family: See **taxonomic rank.**
Feather star: See **crinoid.**
Fisheye lens: A wide-angle camera lens with a field of vision covering up to 180°.
Fission: Reproduction of organisms by division, e.g. in brittle sea stars.

Gastropod: Class of **molluscs** having a single shell and a muscular foot.
Genus: First **taxonomic rank** above species.
Gorgonians: Sessile colonial Cnidaria, commonly known as sea fans and sea whips.
Hard or stony corals: Marine animals of the **order** Scleractinia, in the **phylum Cnidaria** that build calcium carbonate or limestone supporting structures. They are predominantly **colonial** and these forms are referred to as 'reef building corals'; see also **Hexacorallia**. The individuals are called **polyps**.
Herbivore: Plant eater.
Hermaphrodite: Animal having both male and female reproductive organs.
Hexacoral: Hexacorallia is a subclass of **Anthozoa** in the order of **zoantharia**. Their **polyps** generally have a six-fold symmetry, hence the name hexacoral (cf. **octocorals**). The subclass includes all of the **hard** or **stony corals** as well as other animals, e.g. **sea anemones** and **zoantharia.**
Homochromy: A form of camouflage using colour and texture which enables the animal to visually blend into its environment or background.
Hydrodynamic: Branch of physics dealing with the motion of fluids and the resulting forces acting on solid bodies.
Hydroids: Colonial animals in the **phylum Cnidaria** distantly related to sea anemones. The **polyps** are typically quite small, but have potent stinging cells known as **nematocysts.**
Invertebrate: An animal without a backbone.
Isopods: A large **order** of **crustaceans** in the **phylum Arthropoda**. They have a chitinous exoskeleton and jointed limbs.
Iridescent: Exhibiting luminous colours that appear to change when viewed from different angles.
Mantle: The muscle that holds the body of a snail to the shell and may be expanded and retracted depending on the animal's activity.
Medusa: One of the two forms of **Cnidaria** in which the body is umbrella-like or bell-shaped (cf. **polyp**). Plural: medusae.
Metamorphosis: A process in which animals undergo major physical changes after birth. These changes can be quite rapid and extreme involving a complete transformation of body plan.
Mimicry: The ability of an organism to imitate one or more traits or appearance of an unrelated species to create an advantage, e.g. to aid protection or predation. See 'Captivating mimicry' in Chapter 2 and 'Shape shifters' in Chapter 6.
Molluscs: Very large **phylum** of soft-bodied **invertebrates,** which includes **gastropods** (comprising **nudibranchs**), bivalves and **cephalopods**. They usually have shells or their **vestigial** remnants.

Morphology: Branch of life science studying the form and structure of organisms.
Mutualism: See **symbiosis**.
Nematocyst: Coiled thread-like structures that can be fired as a sting typically present in the tentacles of **Cnidaria**.
Notochord: Rudimentary spinal structure found in **chordates**. In vertebrates, the notochord develops into the spinal column.
Nudibranch: A marine **gastropod** without a shell or true gills, but typically has external branching gills at the back, hence the name which means 'naked gills'. The **vestigial** shell is shed after the larval stage.
Obligate: Symbiosis where at least one **species** is unable to survive outside the association, cf. **facultative**.
Octocorals: A subclass of **Anthozoa** with **polyps** generally having an eight-fold symmetry and are generally known as **soft corals** and **gorgonians**, cf. **hard corals** and **stony corals**. Octocorals are also **colonial**, but the **polyps** are embedded in a soft flexible matrix that creates the visible structure of the colony as typically exhibited in sea fans.
Omnivore: An animal that can survive on both plant and animal sources.
Operculum: A lid-like structure that closes over an **aperture** as protection, e.g. in **gastropods** or Christmas tree worms.
Opisthobranch: This is now an informal name for a large and diverse group of **gastropods**.
Order: See **taxonomic rank**.
Oviduct: The tube through which an ovum or egg passes from an ovary.
Ovulid: A **mollusc** belonging to the family Ovulidae (allied to cowries). Also known as false cowries.
Paleozoic: A major interval of geologic time lasting from 541 to 252 million years ago, ending with the Permian extinction, the greatest such event in the Earth's history.
Parasitism: See **symbiosis**.
Parthenogenesis: A natural form of **asexual** reproduction in which the embryos development occur without fertilisation by sperm.
Pelagic: Organisms inhabiting the main **water column** of open oceans and coasts, neither near the bottom or the shore.
Pharyngeal jaws: A supplementary set of jaws located in an animal's throat, or pharynx.
Photoreceptor: A structure in an animal, particularly a sensory cell or organ, that responds to light.
Photosynthesis: A process used by plants and other organisms to convert energy from light into chemical energy, e.g. sugars being synthesised from carbon dioxide and water.

Phytoplankton: Mostly single-celled algae. See **plankton**.
Phylum: See **taxonomic rank**.
Plankton: Minute or microscopic animals (**zooplankton**) or plants (**phytoplankton**) that drift freely within the currents of the oceans.
Piscivore: A carnivore that primarily eats fish.
Polychaetes: A class of segmented worms of the **phylum** Annelida, which includes bristle worms.
Planktivore: An aquatic animal that feeds on **plankton**, including **phytoplankton** and/or **zooplankton**.
Protandry: Relating to an organism that starts life as a male and then changes into a female.
Protean: The ability or tendency to make frequent changes.
Protozoan: An informal term for single-celled microorganisms called **eukaryotes**. They can be free-living or parasitic, and feed on organic matter such as other microorganisms or organic tissues and debris.
Pycnogonida: A **class** of marine **arthropods** known as sea spiders.
Ret mirable: A complex arrangement of arteries and veins lying very close to each other, found in some vertebrates, mostly warm-blooded ones. It utilises counter-current blood flow to act as a heat exchanger.
School: A group of fish that swim together in the same direction in a coordinated and harmonised manner, cf. **shoal**.
Sclerite: A minute, hard, shard-like structure present in some **invertebrates** such as soft corals. Similar structures in soft corals are called **sclerites**, and ossicles in **echinoids**.
Sea anemone: Large solitary **polyps** which have no skeleton. They have a basal disk which they use to fix to the **substrate**.
Sessile: Attached by the base usually to a substrate.
Seta: External bristle or hair (plural setae).
Sexual dimorphism: Relating to the differences between males and females of the same species beyond the sexual organs. It is most evident in external physical appearances, but may also apply to internal organs and biological functions.
Shell aperture: The main opening of the shell of a snail.
Shoal: Aggregation of given **species** of fish having no coordinated movement or direction, cf. **school**.
Soft coral: An **order** of corals that do not produce hard calcium carbonate skeletons. They are **sessile colonial cnidarians** commonly known as **gorgonians** or sea fans. See **octocorals**.
Species: Basic taxonomic rank of classification of a group of organisms which share common characteristics and are capable of interbreeding producing fertile offspring. See **taxonomic rank**.

Species complex: A group of closely related organisms that are so similar in appearance that differentiation between them is difficult.
Spicule: A minute, hard, shard-like structure present in some **invertebrates** such as sponges. Similar structures in soft corals are called **sclerites**, and ossicles in **echinoids.**
Stomatopods: **Order** of marine crustaceans which includes mantis shrimps.
Substrate: The sea bed, including rock, coral, sand or mud, on or in which marine organisms live or are attached.
Symbiont: An organism which is an **associate** of another and usually larger, **species** which is called the host.
Symbiosis: Association between two different **species** living in close physical interaction. There are three main types of symbiosis—**commensalism, mutualism** and **parasitism.** (See "Constructive synergy' in Chapter 7).
Taxonomic rank: In biological classification, taxonomic ranks in descending order of size are—Kingdom, **phylum, class, order, family, genus,** and **species**.
Tectonic: Pertaining to the processes that control the structure and properties of the Earth's crust and its evolution.
Theca: The cup-like central body of a **crinoid**. It has a five-fold symmetry and is analogous to the body or disk of other **echinoids.**

Thermocline: A thin layer in a large body of water with a marked temperature gradient above and below which the water is at different temperatures.
Thoracic: The section of spine in the upper back and abdomen.
Tube feet: Tube feet or podia are small active tubular projections on the oral face of **echinoids** such as sea urchins and sea stars.
Tubercle: See **echinoid**.
Tunicate: See **ascidian**.
Vestigial: The residual part of a body or organ that is typically small or imperfectly developed and not able to function.
Viviparous: The ability to give birth to live young which have been nourished and developed inside the body of the parent.
Water column: Zone of water between the sea bed and the surface.
Zoantharia: An **order** of **Cnidaria** commonly found in coral reefs and includes the **hexacorals,** better known as zoanthids.
Zooplankton: See **plankton**.
Zooxanthellae: Single-celled microscopic algae that are able to live symbiotically with a diverse range of marine invertebrates, including jellyfish, **nudibranchs**, **sponges**, and especially corals.

Photo guide

Front cover: (1) Feather star: *Oxymanthus* sp., arm length 20 cm (2) Hard coral: *Tubastraea micranthus* (3) Titan triggerfish: *Balistoides viridescens*, TL 75 cm (4) Cleaner wrasse: *Labroides dimidiatus*, TL 11.5 cm (5) Red coral grouper: *Cephalopholis miniata*, TL 40 cm. Soft corals: (6) *Melithaea* sp. (7) *Dendronephthya* sp.

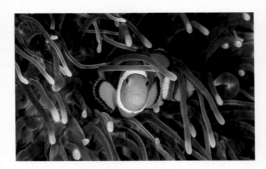

Back cover: Anemone: *Heteractis magnifica* with anemonefish: *Amphiprion ocellaris*, TL 9 cm

Pages ii-iii. (1) Lionfish: *Pterois volitans*, TL (2) Sea fan: *Melithaea* sp. (3) Sea whip: *Ellisella* sp.

Page vii. Contents page (1) Red coral grouper: *Cephalopholis miniata*, TL 40 cm with (3) Shoal of silversides: *Atherinomorus* sp. TL 8 cm (2) Sea fan: *Melithaea* sp.

Page ix. Two-colour parrotfish: *Cetoscarus bicolor*, TL 90 cm between hard coral tables: *Acropora* sp.

Page xi. Lyretail anthias: *Pseudanthias cheirospilos*, TL 10 cm, (1) Male, (2) Female, above (3) Stand of hard coral: *Acropora* sp. (4) Hard coral: *Tubastraea* sp.

Pages xii-1. (1) Damselfish: *Chromis viridis*, TL 10 cm (2) Smoky chromis: *Chromis fumea*, TL 13 cm (3) Hard coral: *Acropora* sp.

Pages 2-3. (1) Side view and (2) Frontal shot of delicate stands of *Montipora* sp. hard coral

Pages 4-5. Reef panorama of hard corals mainly (1) *Acropora* sp. and (2) *Montipora* sp.

Note: This photo guide provides the Latin and common names for the species featured. As the IDs are generally assessed from photographs, they are consequently indicative. A definite ID would normally require study of samples and possibly genetic analysis.

Pages 6-7. Reef panorama of hard corals mainly (1) *Acropora* sp. and (2) *Montipora* sp. with (3) *Isopora palifera* (4) Sergeant damselfish: *Abudefduf vaigiensis*, TL 18 cm

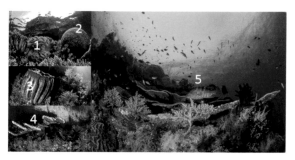

Pages 8-9. (1) Soft coral: *Lemnalia* sp. (2) Hard coral: *Platygyra* sp. (3) Barrel sponge: *Xestospongia testudinaria* (4) Hard coral: *Acropora* sp. and soft coral: *Scleronephthya* sp. (5) Red coral grouper: *Cephalopholis miniata*, TL 40 cm above *Acropora* coral and a variety of soft corals

Pages 10-11. (1) Sea fan: *Melithaea* sp. (2) Ghost goby: *Pleurosicya boldinghi*, TL 3.5 cm in soft coral: *Scleronephthya* sp. (3) See IDs provided for front cover

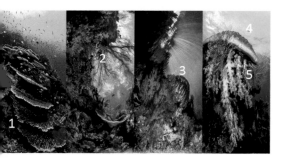

Pages 12-13. (1) Plates of *Acropora* sp. (2) Sea fan: *Melithaea* sp. (3) Sponge: *Xestospongia testudinaria* behind soft coral: *Antipathes* sp. (4) Hard coral: *Porites* sp. (5) *Melithaea* and *Scleronephthya* sp.

Pages 14-15. (1) Fairy basslets: *Pseudanthias squamipinnis* with sponge: *Cribochalina* sp. (2) Clam: *Tridacna* sp. with *Pseudanthias* sp. inside (3) Anemones: Mixture of *Heteractis* and *Entacmea* sp. (4) Hard coral: *Dipsastraea* sp.

Pages 16-17. (1) Parrotfish: *Scarus rubroviolaceus*, (male) TL 70 cm (2) Flounder: *Bothus mancus* with stands of hydrocoral: *Millepora* spp. (3) Bumphead parrotfish: *Bolbometopon muricatum*, TL 120 cm (4) Hard coral: *Isopora palifera* and (5) Hydrocoral: *Millepora* spp.

Pages 18-19. Pinnate batfish: *Platax pinnatus*: (1) juvenile, with cardinalfish: *Ostorhinchus apogonides* TL 10 cm, and (3) adults, TL 37 cm with cleaner wrasse: *Labroides dimidiatus*, TL 11.5 cm and shoal of juvenile convict blennies: *Pholidichthys leucotaenia* (2) Flatworm: *Pseudoceros/Pseudobiceros* sp. TL 7 cm

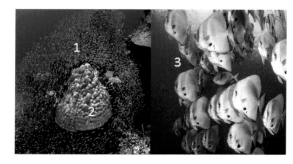

Pages 20-21. (1) Shoal of juvenile convict blennies: *Pholidichthys leucotaenia* (2) *Porites* sp. hard coral (3) Shoal of Blunthead batfish: *Platax teira*, TL 70 cm

Page 22-23. (1) Golden sweepers: *Parapriacanthus ransonneti*, TL 10 cm (2) Black coral: *Antipatharia* sp. (3) Hard coral: *Acropora* sp.

Pages 24-25. All three photos feature aggregations of Ribboned sweetlips: *Plectorhinchus polytaenia*, TL 50 cm (1) with Golden sweepers: *Parapriacanthus ransonneti*, TL 10 cm (2) and (4) Zoanthids: *Parazoanthus* sp. (3) Black coral: *Antipatharia* sp.

Pages 26-27. (1) Bannerfish: *Heniochus diphreutes*, TL 20 cm (2) Sea fan: *Annella* sp. (3) Yellow and blue back fusiliers: *Caesio teres*, TL 40 cm (4) Blue-streak Fusiliers: *Pterocaesio tile*, TL 5 cm

Page 28-29. (1) Blacktip reef shark: *Carcharhinus melanopterus*, TL 180 cm (2) Juvenile Golden trevally: *Gnanthonodon speciosus* (adults reach TL 120 cm)

Page 30-31. (1) Blackfin barracuda: *Sphyraena genie*, TL 140 cm (2) Snub-nose dart: *Trachinotus blochii*, TL 70 cm (3) Blacktip reef shark: *Carcharhinus melanopterus*, TL 180 cm

Page 32-33. Oceanic sunfish: *Mola*, TL 320 cm, being cleaned on left by Butterflyfish: *Chaetodon* sp.

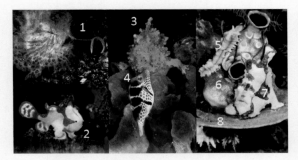

Page 34-35. (1) Hairy frogfish: *Antennarius striatus*, Tl 20 cm (2) and (7) Warty frogfish: *Antennarius maculatus*, TL 11cm (3) Painted frogfish: *Antennarius pictus*, TL 21 cm (4) Black-saddled toby: *Canthigaster Valentine*, TL 10 cm (5) Sea cucumber: *Colochirus robustus*, TL 6 cm (6) Ascidian: *Polycarpa aurata*, TL 8 cm (8) Sponge: *Phyllospongia papyracea*

Page 36-37. (1)–(3) Three-colour variations of the Weedy scorpionfish: *Rhinopias frondosa*, TL 20 cm (4) Fire sea urchin: *Asthenosoma varium*, TL 22 cm

Page 38-39. (1) and (2) Juveniles and (4) Adult of Semicircle angelfish: *Pomacanthus semicirculatus*, TL 35 cm (3) Juvenile splendid dottyback, *Manonichthys splendens*, (adult TL 9 cm) (5) Barrel sponge: *Xestospongia testudinaria*, TL 70 cm (6) Moorish idol: *Zanclus cornutus*, TL 22 cm

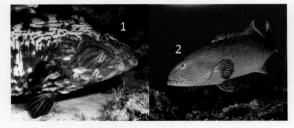

Page 40-41. Highfin coral grouper: *Plectropomus oligacanthus* (1) Nighttime colouration, (2) Daytime colouration TL 75 cm

Pages 42-43. (1) Sea fan: *Melithaea* sp. (2) and (4) Titan triggerfish: *Balistoides viridescens*, TL 75 cm (3) Redtooth triggerfish: *Odonus niger*, TL 40 cm

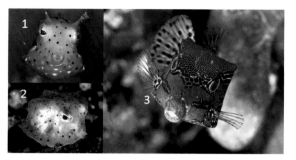

Pages 44-45. (1) and (2) Boxfish juveniles. These are very small (< 2 cm) and hard to ID at this early stage (1) cf. *Lactoria fornasini*, (2) cf. *Ostracion cubicus* (3) Solor boxfish: *Ostracion solorensis*

Pages 46-47. (1) Starry blenny: *Salarias ramosus*, TL 5 cm (2) Springer's blenny: *Cirripectes springeri*, TL 7 cm (3) False cleanerfish: *Aspidontus taeniatus*, TL 10.5 cm with purple encrusting sponge: *Chalinula nematifera* (4) Midas blenny: *Ecsenius midas*, TL 13 cm with golden encrusting sponge

Pages 48-49. Seahorses: (1) *Hippocampus* sp., (4) *H. moluccensis*, TL 16 cm, (8) *H. histrix*, TL 15 cm (2) Leather coral: *Sarcophyton* sp. (3) Feather star: *Comaster* sp. (5) Soft coral: Xeniidae sp. (6) Seamoth: *Eurypegasus draconis*, TL 10 cm (7) Fire sea urchin: *Asthenosoma varium*, TL 22 cm

Pages 50-51. Pygmy seahorses: (1) and (5) *Hippocampus bargibanti*, TL 2 cm on *Muricella* sea fans. Pink colour is more common. (2) *H. pontohi*, TL 1.6 cm (3) *H. denise*, TL 2.4 cm but usually less than 1.6 cm (4) Pygmy pipe dragon: *Kyonemichthys rumengani*, TL 2.4 cm. The TL includes the tail so all of these creatures look really tiny in the field

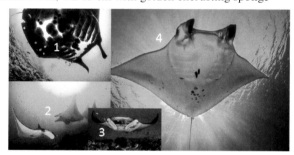

Pages 52-53. (1)–(4) Manta rays: *Manta birostris*, Maximum width 670 cm (1)–(3) at cleaning stations with Butterflyfish: *Chaetodon kleinii*, TL 14 cm acting as cleaners

Page 54-55. (1), (2) and (4) Whale sharks: *Rhincodon typus*, TL 12 metres. Those shown here are all young fish measuring between 6 and 8 metres. (2) Indo-Pacific bottlenose dolphins: *Tursiops aduncus* staying below 20 metres depth below fishing platforms

Pages 56-57. Bigfin reef squid: *Sepioteuthis lessoniana*, TL 36 cm

Pages 58-59. (1) Feather star: *Oxycomanthus* sp. arms up to 26 cm (2) Hard coral: *Acropora* sp. (3) Broadclub cuttlefish: *Sepia latimanus*, TL 50 cm

Page 60-61: (1) and (4) Blue ring octopus: *Hapalochlaena* spp. arm lengths to 7 cm (2) Coralline alga: *Halimeda* sp. (3) and (5) Ocellate or Mototi octopus: *Amphioctopus siamensis*, arm length to 20 cm, showing extreme camouflage in (5)

Pages 62-63. (1) to (3) Textile cone shell: *Conus textile*, TL 9 cm with (2) showing emergence from rubble (4) Bumblebee shrimp: *Gnathophyllum americanum*, TL 1.2 cm

Pages 64-65. (1) and (5) Spiny tiger shrimp: *Phyllognathia ceratophthalmus*, TL 2 cm (2) Leopard anemone shrimp: *Izucaris masudai*, TL 2 cm on (3) Leopard anemone: *Nemanthus nitidus* (4) Bumblebee shrimp: *Gnathophyllum americanum*, TL 1.2 cm

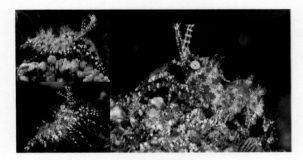

Pages 66-67. The *Saron* or Marbled shrimps are a complex which is not well described and can best be all grouped as *Saron* spp.

Pages 68-69. (1) and (3) Hairy squat lobster: *Lauriea siagiani*, carapace to 1.5 cm. Associates with (4) Giant barrel sponges: *Xestospongia testudinaria*. (2) Sea spider—a marine arthropod of the class: Pycnogonida

Pages: 70-71. (1) Skeleton shrimp: *Caprella* spp. adult female (TL 3 cm) with juveniles. (2) Sea fan: Plexauridae sp. (3) and (5) Hairy or Algae shrimp: *Phycocaris simulans*, up to 6 mm (4) Soft coral: *Paraminabea* sp. (6) Marine algae: *Halymenia* sp.

Pages 72-73. (1) Taylor's garden eel: *Heteroconger taylori*, TL 50 cm (2) Field of garden eels: *Heteroconger* sp. (3) Bobbit worm: *Eunice aphroditois*, TL 100 cm

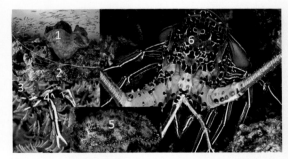

Pages 74-75. (1) Demosponge: *Petrosia* sp. (2) and (6) Painted spiny lobster: *Panulirus versicolor*, carapace to 40 cm (3) Feather star: *Comanthus* sp. (4) Elegant crinoid squat lobster: *Allogalathea elegans*, carapace to 2 cm (5) Sculptured slipper lobster: *Parribacus antarcticus*, TL 20 cm (Note: Despite name, not found in the Antarctic)

Pages 76-77. (1), (2) and (4) Decorator crabs: *Achaeus* sp. 2 cm. (1) and (2) on branching hydroid: *Aglaophenia* sp., transferring hydroid polyps from shed carapace (3) Feather star: *Comanthus* sp.

Pages 78-79. (1) Night sea anemone: *Phyllodiscus semoni*, height when extended 20 cm (2) Luminescent stinging jellyfish, bell size in observed examples around 8 cm (3) Jellyfish: *Mastigias* sp., bell size to 12 cm

Pages 80-81. (1) Tiger cowrie: *Cyprea tigris*, up to 15 cm (2) False cowrie: *Dentiovula colobica*, 1.5 cm on (3) Soft coral: *Acanthogorgia* sp. (4) Opisthobranch: *Micromelo undatus*, shell length up to 1.7 cm (5) False cowrie: *Serratovolva dondani*, shell size to 2 cm, on soft coral: *Dendronephthya* sp.

Pages 82-83. (1) Giant frogfish: *Antennarius commerson*, TL 45 cm Ascidians: (2) *Clavelina* sp., (3) Lollipop tunicate: *Nephtheis fascicularis*, (4) *Rhopalaea* sp.

Pages 84-85. (1) Blackfin barracuda: *Sphyraena qenie*, TL 140 cm (2) Silversides: *Atherinomorus* sp. TL 8 cm

Page 86-87. (1) Redfin anthias: *Pseudanthias dispar*, TL 9.5 cm, male with 4 females (2) Square-spot anthias: *Pseudanthias pleurotaenia*, TL 20 cm, male (3) Blue-streak fusiliers: *Pterocaesio tile*, TL 25 cm (4) Barrel sponge: *Xestospongia testudinaria*

Page 88-89. Mixed shoals of *Pseudanthias* sp. above (2) Feather stars: *Comanthus* sp. (arm length 20 cm) (3) Sponge: Niphatidae sp. and (4) Hard coral: *Acropora* sp.

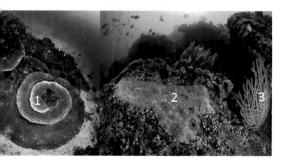

Pages 90-91. (1) Juvenile wobbegong shark sitting on hard coral *Montipora* sp. (2) Tasselled wobbegong: *Eucrossorhinus dasypogon*, TL typically 180 cm, possibly up to 360 cm (3) Sea fan: *Euplexaura* sp.

Page 92-93. (1) Bigeye trevally: *Caranx sexfasciatus*, TL 85 cm (2) Giant trevally: *Caranx ignobilis*, TL 165 cm

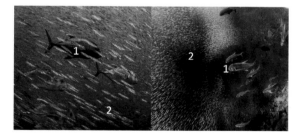

Page 94-95. (1) Giant trevally: *Caranx ignobilis*, TL 165 cm (2) Silversides: *Atherinomorus* sp. TL 8 cm

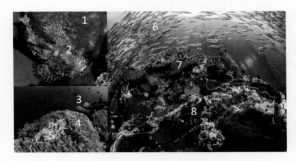

Pages 96-97. (1) Demosponge (2) Scorpionfish: *Scorpaenopsis oxycephala*, TL 35 cm (3) Redfin anthias: *Pseudanthias dispar*, TL 9.5 cm (4) Soft coral: Xeniidae sp. (5) as (2) and (8). (6) Fusiliers: *Pterocaesio* sp. (7) Soft coral: Alcyoniidae sp.

Pages 98-99. Reef lizardfish: *Synodus variegatus*, TL 28 cm (2) Slender lizardfish: *Saurida gracilis*, TL 28 cm. (Note: This differs from *Synodus* sp. as jaw teeth are visible when the mouth is closed.) (3) Star gazer: *Uranoscopus* sp. TL 35 cm (4) Shrimp goby: *Cryptocentrus* sp. TL 10 cm

Pages 100-101. (1) Napoleon wrasse: *Cheilinus undulatus*, TL 2 metres (2) Trevally: *Carangoides* sp. (3) Yellow goatfish: *Mulloidichthys vanicolensis*, TL 30 cm (4) Trumpet fish: *Aulostomus chinensis*, TL 60 cm (5) Blue-spotted stingray: *Taeniura lymma*, TL 70 cm (6) Soft coral: *Litophyton* sp.

Page 102-103. (1) Fimbriated moray eel: *Gymnothorax fimbriatus*, TL 80 cm (2) X-rays showing jaw moray jaw Bones (courtesy of Dr. R. Mehta). (3) Black-spotted moray: *Gymnothorax favagineus*, TL 180 cm (4) Cardinalfish: *Ostorhinchus apogonides*, TL 10 cm

Page 104-105. (1) Sea anemone: *Stichodactyla* sp. (2) Porcelain crab: *Neopetrolisthes maculatus*, carapace to 3 cm. (3) Porcelain crab: *Lissoporcellana nakasonei*, 1 cm. (4) Sea pen: *Virgularia* sp. (5) Barnacle: *Nobia grandis* in hard coral: *Galaxea* sp. (6) Haig's porcelain crab: *Porcellana haigae*, 2 cm. (7) Soft coral: Xeniidae sp.

Pages 106-107 (1) Harlequin shrimp: *Hymenocera picta*, 5 cm, female with smaller male (2) Blue sea star: *Linckia laevigata*, 30 cm (3) Multipore sea star: *Linckia multifora* regenerating arms (4) *H. picta* with juvenile (6), (5) Severed leg of sea star: *Formia* sp.

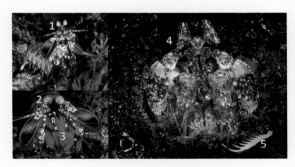

Pages 108-109. (1) Peacock mantis shrimp: *Odontodactylus scyllarus*, 18 cm, (2) with eggs (3), (4) Giant mantis shrimp: *Lysiosquillina lisa* (5) Lower half of spear from previous shed of *L. lisa*

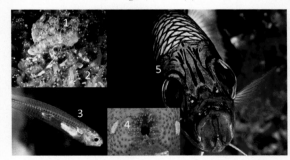

Page 110-111. (1) Hermit crab: Diogenidae sp., at night walking over extended polyps of hard coral. (3) Silver pearlfish: *Encheliophis homei*, TL 19 cm (4) Leopard sea cucumber: *Bohadschia argus* with juvenile brittle stars, *Ophiothela* sp., around anus (5) Shadowfin soldierfish: *Myripristis adusta*, TL 15 cm

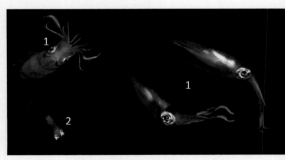

Page 112-113, (1) Bigfin reef squid: *Septioteuthis lessoniana*, TL 36 cm. These were encountered at night and were quite small, around 12 cm. (2) Squid embryo developing in egg case

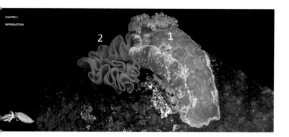

Pages 114-115. (1) Spanish dancer nudibranch: *Hexabranchus sanguineus*, TL 40 cm with eggs (2)

Pages 116-117. Nudibranchs: (1) *Hypselodoris bullocki*, 4.5 cm (2) *Tambja* sp., 6 cm; *Nembrotha chamberlaini*, 6 cm (3) Laying eggs while consuming colonial ascidians; and (5) Mating in ménage à trois (4) Associate shrimp: *Zenopontonia rex*

Pages 118-119. Nudibranchs: Mass spawning of *Gymnodoris ceylonica*, 5 cm

Pages 120-121. (1) Sea pen: *Virgularia* sp. (2) Brittle star: *Ophiothrix* sp. and juveniles (4), (3) Soft coral: *Clavularia* sp. (5) Brittle stars mating: *Ophiothrix purpurea*, arm length to 15 cm (6) Leathery soft coral: Alcyoniidae sp.

Pages 122-123. (1) Tiger anemone or Gorgonian wrapper: *Nemanthus annamensis*, individuals up to 4 cm (2) *N. annamensis* colour variation (3) Sea whip: Ellisellidae sp. (4) Common whip goby: *Bryaninops yongei*, TL 3.7 cm

Pages 124-125. (1) Banggai cardinalfish: *Pterapogon kauderni*, TL 7 cm (2) Juvenile *P. kauderni* in Radiant sea urchin: *Astropyga radiata*, 20 cm, and (3) With *Janolus* sp. nudibranch to 4 cm (4) *P. kauderni* surrounding (5) Sea anemone: *Heteractis magnifica*, 40 cm disc width (6) Pink anemonefish: *Amphiprion perideraion*, TL 8 cm

Pages 126-127. (1) Pufferfish: *Canthigaster bennetti*, TL 9 cm, dense aggregations and (2) close-up (3) Anemonefish: *Amphiprion clarkii*, TL 12 cm (4) Sea anemone: *Heteractis magnifica*. Top left photo courtesy of Andrew Podzorski.

Pages 128-129. (1) Ornate ghost pipefish, *Solenostomus paradoxus*, TL 11 cm, close-up with eggs (3) Same species, full profile, eggs also visible; both females (2) Yellow banded pipefish: *Dunckerocampus pessuliferus*, TL 16 cm, male with eggs.

Pages 130-131. Mandarinfish: *Pterosynchiropus splendidus*, TL 6 cm (1) Close-up of a male, (2) Mating pair (3) Sperm and eggs being broadcast

Pages 132-133. (1) and (7) Flasher wrasse: *Paracheilinus nursalim* males in mating display (note variation in number of dorsal fin rays), and (6) Female. (2) *Paracheilinus filamentosus*, male displaying. (4) and (5) Possibly Walton's flasher wrasse: *Paracheilinus walton*, TL 6 cm, young males displaying.

Pages 134-135. (1) Orange spotted trevally: *Carangoides bajad*, TL 55 cm, and (3) Showing rapid change to yellow phase. (2) Silversides: *Atherinomorus* sp. TL 8 cm (4) Sea fans: *Melithaea* sp.

Pages 136-137. (1) and (3) Blackfin barracuda: *Sphyraena genie*, TL 140 cm (2) Oxeye scad: *Selar boops*, TL 26 cm

Pages 138-139. (1) Giant trevally: *Caranx ignobilis*, TL 165 cm, shadowing manta ray and on right 3 pursuing to join (2) Manta rays: *Manta birostris*, maximum width 670 cm (3) Three juvenile Golden trevally: *Gnanthonodon speciosus*, (adults reach TL 120 cm)

Pages 140-141. (1) Round batfish: *Platax orbicularis*, four juveniles with floating leaf, and two below showing colour variation (adults reach TL 50 cm) (2) Blunthead batfish: *Platax teira*, juvenile (adults reach TL 70 cm) (3) Four juvenile *P. orbicularis* with juvenile Silversides

Pages 142-143. Round batfish: *Platax orbicularis*, juveniles showing further colour variations

Pages 144-145. (1) Taylor's inflator filefish: *Brachaluteres taylori*, tiny juvenile (adults max 5 cm), in soft coral: Xeniidae sp. (2) Filefish juvenile: *Brachaluteres* sp. (3) Feather dusty worm: Sabellidae sp. (4) Radial filefish: *Acreichthys radiatus*, TL 7 cm (5) Sea fan: *Siphonogorgia* sp.

Pages 146-147. (1) Striped sweetlips: *Plectorhinchus lessonii*, juvenile by feather star: Crinoidae sp.
(2) Harlequin sweetlips: *Plectorhinchus chaetodonoides*, juvenile (side view shown in lower photo on page xxx)
(3) Humpback grouper: *Chromileptes altivelis*, juvenile

Pages 148-149. Mimic octopus: *Thaumoctopus mimicus*, arm length to 30 cm (1) Surveying environment from burrow. Tentative mimicry subjects: (2) Sea snake (3) Flounder and (4) Lionfish or toxic anemone

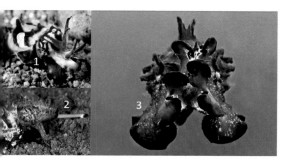

Pages 150-151. Flamboyant cuttlefish: *Metasepia pfefferi*, up to 8 cm (1) Bright aposematic colouration when excited (2) Subdued tones and extending feeding tentacle (3) Unusual behaviour rising high in water column

Pages 152-153. Veined or coconut octopus: *Amphioctopus marginatus*, arms up to 30 cm with coconut shells

Pages 154-155. Veined or coconut octopus: *Amphioctopus marginatus*, arms up to 30 cm utilising (1) Beer bottle, (2) Clear broken bottle, and (3) Grey bonnet shell: *Phalium glaucum* (15 cm) and green broken bottle

Pages 156-157. Bobtail squid: *Euprymna berryi*, body to 5cm (1) Hovering over substrate, (2) Emerging from substrate, (3) Burying itself in substrate

Pages 158-159. (1) Decorator crab on sponge: *Xestospongia testudinaria* (2) Colonial ascidians: *Clavelina* sp. (3) Feather hydroid: cf. *Aglaophenia* sp. (4) Sponge crab: cf. juvenile *Dromia dormia* (adult to 20 cm) with sponge: *Clathria reinwardti* (5) Corallimorph decorator crab: *Cyclocoeloma tuberculata* (carapace 5 cm) (7) Corallimorpharians and (8) soft corals. (6) Ascidians: *Didemnum molle*, 2 cm

Pages 160-161. (1) Boxer crab: *Lybia tessellate* with eggs holding sea anemone of Aliciidae family cf. *Triactis producta* (2) Hermit crab: *Dardenus* sp., cf. *D. gemmatus* with anemones (3): cf. *Calliactis polypus*. (4) Teddy bear crab: *Polydectus cupulifer* with pair of anemones (5): *Phellia* sp.

Pages 162-163. Longnose hawkfish: *Oxycirrhites typus*, TL 8 cm (2) Sea fan: *Chironephthya* sp.

Pages 164-165. (1) Randall's shrimp goby: *Amblyeleotris randalli*, Tl 11 cm (2) Goby shrimp: *Alpheus* sp. (3) Threadfin shrimp goby: *Stonogobiops nematodes*, TL 6 cm

Pages 166-167. (1) Clownfish: *Amphiprion percula*, TL 7.5 cm (2) False clownfish: *Amphiprion ocellaris*, TL 9.5 cm (3) Spinecheek clown fish: *Premnas biaculeatus*, TL 12 cm. All fish shown in sea anemone: *Heteractis magnifica*, disc size 40 cm

Pages 168-169. (1) Clownfish: *Amphiprion percula*, TL 7.5 cm (2) False clownfish: *Amphiprion ocellaris*, TL 9.5 cm (3) Juvenile Banggai cardinalfish: *Pterapogon kauderni* (4) Blue-streak fusiliers: *Pterocaesio tile*, TL 25 cm (5) Pink anemonefish: *Amphiprion perideraion*, TL 10 cm All sea anemones: *Heteractis magnifca*, disk size 40 cm

Pages 170-171. Zebra crab: *Zebrida adamsii*, (1) Showing cryptic carapace and (2) With eggs (3) Fire sea urchin: *Asthenosoma varium*

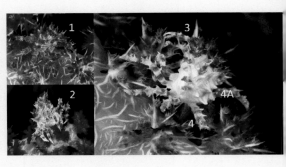

Pages 172-173. (1) to (3) Colour variations in Candy crabs: *Hoplophrys oatseii* (the white one is omitted for clarity) on soft coral: *Dendronephthya* sp. (4) Sea urchin larval form with (4A) showing paddle-like appendage

Pages 174-175. (1) Gall crab: *Lithoscaptus* sp. (2) Hermit crab: *Paguritta morgani* in hard coral *Montipora* sp. (3) Christmas tree worm: *Spirobranchus* sp. (4) Scallop: *Pedum spondyloidem*

Pages 176-177. (1) Crinoid shrimp: *Laomenes* sp. (2) Striped snapping shrimp: *Synalpheus striatus* sitting in the centre of the crinoid on the theca (3) Crinoid clingfish: *Discotrema crinophila*, TL 3 cm with (4) Showing thoracic sucking disk

Pages xx-xx. (1) White-spotted puffer: *Arothron hispidus*, TL 48 cm (2) Cheek-lined wrasse: *Oxycheilinus digramma* TL 30 cm (3) Oriental wrasse: *Oxycheilinus orientalis*, TL 18 cm All with Blue-streak cleaner wrasse: *Labroides dimidiatus*, TL 11.5 cm

Pages 180-181. (1) Yellowtail fusilier: *Caesio cuning*, TL 50 cm in cleaning colouration with (2) Blue-streak cleaner wrasse: *Labroides dimidiatus*, TL 11.5 cm (3) Mottled moray: *Gymnothorax pseudothyrsoideus*, TL 80 cm (4) Clear cleaner shrimp: *Urocaridella antonbrunni*, 3 cm

Pages 182-183. (1) Whitetip reef shark: *Triaenodon obesus*, TL 170 cm (2) Yellowtail fusilier: *Caesio cuning*, TL 50 cm (3) Mixed shoals of fusiliers (4) Giant trevally: *Caranx ignobilis*, TL 165 cm (5) Palefin unicornfish: *Naso brevirostris*, TL 50 cm (6) Surgeonfish: *Naso* sp. (7) Blue-streak cleaner wrasse: *Labroides dimidiatus*, TL 11.5 cm

Pages 184-185. (1) and (2) Painted frogfish: *Antennarius pictus*, TL 16 cm with lure extended (3) Feather duster worm: *Sabellastarte* sp. approx. size 7 cm (4) Frogfish placing pectoral fin 'foot' on worm

Pages 186-187. (1) Whale shark: *Rhincodon typus*, TL 12 metres with (2) Remoras or sharksuckers: large aggregation of Common remoras: *Remora*, TL 60 cm, with some striped Slender sharksuckers: *Echeneis naucrates*, TL 1 metre (3) Pelvic fins—absence of claspers confirming this individual as a female

Pages 188-189. Reef panorama with green sea turtle: *Chelonia mydas*, up to 150 cm

Pages 190-191. Golden sea snake: *Aipysurus laevis*, TL 2 metres: (1) Returning from surface, (2) Hunting for prey around sea anemone, and (4) Breathing at surface (3) False clown fish: *Amphiprion ocellaris*, TL 9.5 cm with Bleached sea anemone: *Heteractis magnifica*

Pages 192-193. Komodo dragons: *Varanus komodiensis*, TL 3 metres

Pages 194-195. Hawksbill turtle: *Eretmochelys imbricata*, up to 1 metre: (1) Feeding amongst *Acropora* coral, (2) Consuming demosponge, (3) Petrosiidae sp. (4) Sitting on reef amongst sponges: (5) *Petrosia* sp. and (6) *Gelloides fibulata* with feather hydroids to right

Pages 196-197. (1) Green sea turtle: *Chelonia mydas*, up to 150 cm (2) Sea fan: *Isis hippuris* (3) Demosponge (4) Soft coral: *Tubipora musica* (5) Manta ray: *Manta birostris*, maximum width 670 cm

Pages 198-199. (1) Underside of table of *Acropora* hard coral (2) Group of five whitetip reef sharks: *Triaenodon obesus*, (adults up to TL 170 cm)

Pages 200-201. Mauwara, Fak, West Papua. Photo courtesy Bruno Hopff, Amira

Pages 202-203. Close-ups of hard coral: *Diploastrea heliopora*: (1) Polyps partially bleached, (2) Fully bleached, (3) Healthy unbleached (4) Pink anemonefish: *Amphiprion perideraion*, TL 10 cm, with bleached sea anemone: *Heteractis magnifca*, disk size 40 cm. (5) Unbleached *Heteractis magnifica* (6) Bleached hard coral: *Isopora palifera*

Pages 204-205. (1) Dead hard coral being overgrown by algae and sponges (2) Crown of thorns sea star: *Acanthaster planci*, 30 cm. Bleached mushroom corals: (3) *Herpolitha limax* and (4) *Pleuractis* sp. Hard corals devastated by blast fishing: (5) *Platygyra* sp. surrounded by (6) *Acropora* sp.

Pages 206-207. (1) Dorid nudibranch: *Hypselodoris emma*, 4 cm (2) Pair of Ribbon moray eels: *Rhinomuraena quaesita*, TL 1.2 metres (3) Juvenile Broadclub cuttlefish: *Sepia latimanus* (adults up to 50 cm)

Pages 208-209. Scorpion lionfish: (1) *Pterois miles*, TL 31 cm and (3) *Pterois volitans*, TL 38 cm (2) Hard coral *Tubastraea micranthus* (4) Sea fan: *Melithaea* sp.

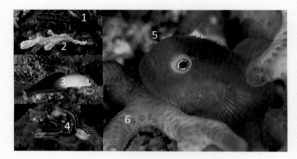

Pages 210-211. (1) Coralline algae (2) Halimeda ghost pipefish: *Solenostomus halimeda*, TL 6.5 cm (3) Bath's coral blenny: *Escenius bathi*, TL 4.4 cm (4) Elegant dartfish: *Nemateleotris decora*, TL 8.5 cm (5) Golden coral goby: *Paragobiodon xanthosoma*, TL 4 cm, on (6) Hard coral: *Seriatopora/Stylophora* sp.

Pages 212-213. (1) Hard coral: *Acropora* sp. (2) Hard coral: *Oulophyllia* sp. (3) Volcanic ash flow above and below water (4) Sea fans: *Melithaea* sp. (5) Hard coral: *Porites* sp.

Page 214. Penemu Island, Raja Ampat, West Papua

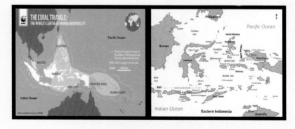

Pages 216-217. Maps

Further reading

General

Allen G.R. and Erdmann M.V. (2012) Reef fishes of the East Indies, Volumes I-III. Tropical Reef Research, Perth, Australia.

Allen G.R. and Steene R. (2002) Indo-Pacific coral reef field guide. Tropical Reef Research, Singapore.

Colin P.L. and Arneson C. (1995) Tropical Pacific invertebrates. Coral Reef Press, USA.

Debelius H. (2001) Asia Pacific reef guide. IKAN-Unterwasserarchiv, Frankfurt, Germany.

Debelius H. (2001) Crustacea. IKAN-Unterwasserarchiv, Frankfurt, Germany.

Fabricius K. and Alderslade P. (2001) Soft corals and sea fans. Australian Institute of Marine Science and the Museum and Art Gallery of the Northern Territory, Australia.

Ferrari A. and Ferrari A. (2003) A diver's guide to underwater Malaysia. Macrolife, Nautilus Publishing, Malaysia.

Gosliner T.M., Behrens D.W. and Williams G.C. (1996) Coral reef animals of the Indo-Pacific. Sea Challengers, USA.

Humann P. and Deloach N. (2010) Reef creature identification. New World Publications Inc., USA.

Jones B. and Shimlock M. (2011) Diving Indonesia's Bird's Head seascape. Saritaksu Editions, Bali with Conservation International Indonesia.

Kassem K. and Madeja E. (2014) The Coral Triangle. John Beaufoy Publishing, Oxford, UK.

Kuiter R.H. and Debelius H. (2009) World atlas of marine fauna. IKAN-Unterwasserarchiv, Frankfurt, Germany.

Ryanskiy A.S. (2018) Coral reef crustaceans from Red Sea to Papua. Reef ID books.

Chapter 1

Erhardt H. and Knop D. (2005) Corals: Indo-pacific field guide. IKAN-Unterwasserarchiv, Frankfurt, Germany.

Huang D., Goldberg E.E., Chou L.M. and Roy K. (2018) The origin and evolution of coral species richness in a marine biodiversity hotspot. Evolution 72: 288–302.

Keith S.A., Baird A.S.A., Hughes T.P., Madin J.S. and Connolly S.R. (2013) Faunal breaks and species composition of Indo-Pacific corals: the role of plate tectonics, environment and habitat distribution. Proceedings of the Royal Society B: Biological Sciences 280: 20130818.

Larsson M. (2012) Why do fish school? Current Zoology 58: 116–128.

Newman L. and Cannon L. (2003) Marine flatworms: The world of polyclads. CSIRO publishing, Australia.

Sheppard A. (2015) Coral reefs: Secret cities of the seas. Natural History Museum, London, UK.

Sheppard R.C., Davy S.K. and Pilling G.M. (2009) The biology of coral reefs. Oxford University Press, UK.

Veron J.E.N., Devantier L.M., Turak E., Green A.L., Kininmonth S., Stafford-Smith M. and Peterson N. (2009) Delineating the Coral Triangle. Galaxea, Journal of Coral Reef Studies 11: 91–100.

Chapter 2

Balcombe J. (2016) What a fish knows: the inner lives of our underwater cousins. Oneworld Publications, UK.

Britz R. and Johnson D.G. (2005) Leis' conundrum: homology of the clavus of the ocean sunfishes. 1. Ontogeny of the median fins and axial skeleton of *Monotrete leiurus* (Teleostei, Tetraodontiformes, Tetraodontidae). Journal of Morphology 266:1–10.

Champ C.M., Vorobyev M. and Marshall N.J. (2016) Colour thresholds in a coral reef fish. The Royal Society, R Soc Open Sci 3: 160399, 1–9.

Gill S. (2003) Blood in the sea: HMS Dunedin and the enigma code. Weidenfeld and Nicolson, London, UK.

Gilman C. (2015) Boxfish don't swim the straight and narrow. Journal of Experimental Biology 218: 1620.

Gomon M.F. (2007) A new genus and miniature species of pipehorse (Syngnathidae) from Indonesia. Aqua, International Journal of Ichthyology 13: 25–30.

Hove J.R., Bryan L.M., Gordon M.S., Webb P.W. and Weihs D. (2001) Boxfishes (Teleostei: Ostraciidae) as a model system for fishes swimming with many fins: kinematics. Journal of Experimental Biology 204: 1459–1471.

Johnson D.G. and Britz R. (2005) Leis' conundrum: homology of the clavus of the ocean sunfishes. 2. Ontogeny of the median fins and axial skeleton of *Ranzania laevis* (Teleostei, Tetraodontiformes, Molidae). Journal of Morphology 266: 11–21.

Joung S-J., Chen C-T., Clark E., Uchida S. and Haung Y.P. (1966) The whale shark, *Rhincodon typus*, is a livebearer: 300 embryos found in one 'megamamma' supreme. Environmental Biology of Fishes 46: 219–223.

Kelley J. (2020) Dazzling or deceptive? The markings of coral reef fish, http://theconversation.com.

Kuiter R.H. (2003) Seahorses, pipefishes and their relatives. TMC Publishing, UK.

Losey G.S. (1972) Predation protection in the poison-fang blenny, *Meiacanthus atrodorsalis*, and its mimics, *Ecsenius bicolor* and *Runula laudandus* (Blenniidae). Pacific Science 26: 129–139.
Price A.C., Weadick C.J., Shim J. and Rodd F.H. (2008) Pigments, patterns, and fish behavior. Zebrafish 5: 297–307.
Randall J.E. (2005) A review of mimicry in marine fishes. Zoological Studies 44: 299–328.
Sharfman W. (2006) Mercedes and the boxfish. The Scientist 20: 17.
Thompson D.W. (1917) On growth and form. Cambridge University Press, UK.
Vincent A.C.J., Foster S.J. and Koldewey H.J. (2011) Conservation and management of seahorses and other Syngnathidae. Journal of Fish Biology 78: 1681–1724.
Wilson A.B. and Orr J.W. (2011) The evolutionary origins of Syngnathidae: pipefishes and seahorses. Journal of Fish Biology 78: 1603–1623.
Zuberbühler T. (2018) Frogfishes: Southeast Asia, Maldives and Red Sea, http://www.critter.ch/frogfish-book.html.

Chapter 3

Hanlon T.H. and Messenger J.B. (1996) Cephalopod behaviour. Cambridge University Press, UK.
Locke A. and Carman M. (2009) An overview of the 2nd international invasive sea squirt conference: what we learned. Aquatic Invasions 4: 1–4.
Lorenz F. and Fehse D. (2009) The living Ovulidae. Conchbooks, Hackenheim, Germany.
Norman M. (2003) Cephalopods: a world guide. Conchbooks, Hackenheim, Germany.
Satoh N. (2014) Developmental genomics of ascidians. Wiley-Blackwell, UK.
Wicksten M.K. (1980) Decorator crabs. Scientific American 2442: 146–157.

Chapter 4

Abbott A. (2015) Octopus genome holds clues to uncanny intelligence. Nature News, doi: 10.1038/nature.2015.18177.
Crane R.L., Cox S.M., Kisare S.A. and Patek S.N. (2018) Smashing mantis shrimp strategically impact shells. Journal of Experimental Biology 221: jeb176099.
DeVries M.S., Murphy E.A.K. and Patek S.N. (2012) Strike mechanics of an ambush predator: the spearing mantis shrimp. Journal of Experimental Biology 215: 4374–4384.
Mehta R.S. and Wainwright P.C. (2007) Raptorial jaws in the throat help moray eels swallow large prey. Nature 449: 79–82.

Chapter 5

Allen G.R., Erdmann M.V. and Yusmalinda N.L.A. (2016) Review of the Indo-Pacific flasherwrasses of the genus *Paracheilinus* (Perciformes: Labridae), with descriptions of three new species. Journal of the Ocean Science Foundation 19: 18–90.
Behrens D.W. (2005) Nudibranch behaviour. New World Publications Inc., USA.
Coleman N. (2008) Nudibranchs encyclopedia. Neville Coleman's Underwater Geography Pty. Ltd., Springwood, Qld., Australia.
Gosliner T.M., Behrens D.W. and Valdés Á. (2008) Indo-Pacific nudibranchs and sea slugs. Sea Challengers and the California Academy of Sciences, USA.
Goslinger T.M., Valdés Á. and Behrens D.W. (2015) Nudibranch & sea slug identification. New World Publications Inc., USA.
Ryansiky A. and Ivanov Y. (2020) Nudibranchs of the Coral Triangle. Reef ID Books.
Vagelli A.A. (2011) The Banggai Cardinalfish: natural history, conservation and culture of *Pterapogon kauderni*. Wiley-Blackwell, UK.
Vail A.L. and Sinclair-Taylor T. (2011) Mass schooling and mortality of *Canthigaster bennetti* in Sulawesi. Coral Reefs 30: 251.
Whitfield J. (2001) Eyes in their stars: Engineers envy brittle star bones' built-in lenses. Nature News, doi: 10.1038/news010823-11.

Chapter 6

Dean S. (2019) Octopus and squid evolution is officially weirder than we could have ever imagined. https://www.sciencealert.com.
Fin J.K., Tregenza T. and Norman M.D. (2009) Defensive tool use in a coconut carrying octopus. Current Biology 19: R1069–1070.
Norman M.D., Finn J. and Tregenza T. (2001) Dynamic mimicry in an Indo-Malayan octopus. Proceedings of the Royal Society B: Biological Sciences 268: 1755–1758.
Schnytzer Y., Gimana Y., Karplus I. and Achituv Y. (2013) Bonsai anemones: growth suppression of sea anemones by their associated kleptoparasitic boxer crab. Journal of Experimental Marine Biology and Ecology 448: 265–270.

Chapter 7

Fautin D.G. and Allen G.R. (1994) Anemone fishes and their host sea anemones. Tetra Press, Germany.
Jaafar Z. and Zeng Y. (2012) Visual acuity of the goby-associated shrimp, *Alpheus rapax* Fabricius, 1798 (Decapoda, Alpheidae). Crustaceana 85: 1487–1497.
Liu J.C.W., Høeg J.T. and Chan B.K.K. (2016) How do coral barnacles start their life in their hosts? The Royal Society Biology Letters 12: 20160124.
Losey G.R. (1979) Fish cleaning symbiosis: proximate causes of host behaviour. Animal Behavior 27: 669–685.
Losey G.R. (1987) Cleaning symbiosis. Symbiosis 4: 229–258.

Chapter 8

Ciofi C., Beaumont M.A., Swingland I.R. and Bruford M.W. (1999) Genetic divergence and units for conservation in the Komodo dragon *Varanus komodoensis*. Proceedings of the Royal Society B: Biological Sciences 266: 2269–2274.

Crowe-Riddell J.M., D'Anastasi B.R., Nankivell J.H., Rasmussen A.R. and Sanders K. (2019) First records of sea snakes (Elapidae: Hydrophiinae) diving to the mesopelagic zone (>200 m). Austral Ecology 44: 752–754.

Hocknull S.A., Piper P.J., van den Bergh G.D., Due R.A., Morwood M.J. and Kurniawan I. (2009) Dragon's paradise lost: palaeobiogeography, evolution and extinction of the largest-ever terrestrial lizards (Varanidae). PLoS One 4: e7241.

NOAA Fisheries: Green Turtle (2020):https://www.fisheries.noaa.gov.

Spotila J.R. (2005) Sea turtles: a complete guide to their biology, behavior, and conservation. Johns Hopkins University Press, USA.

WWF: Green Turtle (2020): https://www.worldwildlife.org/species/green-turtle.

WWF: Hawksbill Turtle (2020): https://www.worldwildlife.org/species/hawksbill-turtle.

Chapter 9

Albins M.A. and Hixon M.A. (2011) Worst case scenario: potential long-term effects of invasive predatory lionfish (*Pterois volitans*) on Atlantic and Caribbean coral-reef communities. Environmental Biology of Fishes 96: 1–7.

Brandl S.J., Tornabene L., Goatley C.H.R., Casey J.M., Morais R.A., Côté I.M., Baldwin C.C., Parravicini V., Schiettekatte N.M.D. and Bellwood D.R. (2019) Demographic dynamics of the smallest marine vertebrates fuel coral reef ecosystem functioning. Science 364: 1189–1192.

Braverman I. (2018) Coral whisperers. University of California Press, USA.

Burke L., Reytar K., Spalding M. and Perry A. (2012) Reefs at risk revisited in the Coral Triangle. World Resources Institute.

Darwin C. (1842) The structure and distribution of coral reefs. Quarterly Journal of the Geological Society 1845: 381–389.

Eakin C.M., Sweatman H.P.A. and Brainard R.E. (2019) The 2014–2017 global-scale coral bleaching event: insights and impacts. Coral Reefs 38: 539–545.

Le Page M. (2019) The tiniest fish are the most important for healthy coral reefs. New Scientist.

Pet-Soede L. and Erdmann M.V. (1998) An overview and comparison of destructive fishing practices in Indonesia. SPC Live Reef Fish Information Bulletin #4.

Roberts C. (2012) The ocean of life: the fate of man and the sea. Viking, Penguin Group, New York.

Rosen B.R. (1982) Darwin, coral reefs and global geology. BioScience 32: 519–525.

Suggett D.J. and Smith D.J. (2011) Interpreting the sign of coral bleaching as friend vs. foe. Global Change Biology 17: 45–55.

Vincent A.C.J., Sadovy de Mitcheson Y.J., Fowler S.L. and Lieberman S. (2013) The role of CITES in the conservation of marine fishes subject to international trade. Fish and Fisheries 15: 563–592.

Index

Alor, 14
Ambon, 18, 106, 206
Ambush predator, 34, 36, 72, 90, 96, 98, 100, 184
Anchovies, 48, 54, 92
Anemone, 104, 122, 124, 148, 158, 160, 166, 190, 194, 202
   Night sea, 78
   Tiger, 64, 122
Anemone city, 14, 166
Anemonefish, 14, 104, 124, 126, 166, 190
Angelfish, 38, 82, 210
Anthia, 14, 86, 176
Aposematic, 60, 80, 106, 130, 148, 150, 170
Aristotle, 32, 82
Ascidian, 82, 116, 144, 158
   life cycle, 82

Bacteria, 156, 192
   bioluminescent, 156
Barnacle, 104
Barracuda, 26, 30, 42, 92, 100, 136, 170
Blenny, 30, 46, 104, 178, 210
   Convict, 18
   Fang, 46, 178
   Midas, 46
Biodiversity, 2, 8, 10, 202, 208, 212
Boxfish, 44, 82
Bryozoan, 144
Butterflyfish, 32, 52, 166, 178

Camouflage, 36, 38, 60, 76, 80, 90, 96, 98, 100, 106, 144, 146, 148, 150, 152, 156, 176, 184
Cardinalfish, 18
   Banggai, 124, 208
   Caribbean, 126, 208
Chagos archipelago, 8, 202

Clam, 14
Chromatophore, 38, 112
Cleaning, Cleaner fish, 18, 38, 46, 138, 88, 182
   False cleaner fish (see fang blenny)
   Cleaner shrimp, 178, 181
   station, 18, 32, 42, 46, 52, 138, 178, 182
Climate change, 8, 50, 122, 126, 176, 202, 204, 208, 210
Clingfish, 176
Cnidarian, 68, 104, 158, 160, 202
Conservation 26, 26, 184, 194, 196, 198, 202, 219
Coral, hard (hexacoral), 2, 78, 90, 96, 110, 104, 158, 174, 202, 212, 219, 220
   soft (octocoral), 10, 70, 80, 96, 104, 158, 172, 202, 218, 220, 221
Coral polyps, 10, 50, 70, 80, 110, 122, 144, 172, 202
Coral bleaching, 8, 10, 126, 190, 202
Coral reefs, beauty of, viii, x, 2, 8, 10
   evolution of, x, 2, 10, 212
   shallow reefs, 2–8, 124, 202
Cowfish – see Boxfish
Crab, Boxer, 160
   Candy, 172
   Decorator, 76, 106, 158
   Fairy, 68
   Gall, 174
   Hermit, 74, 110, 152, 160, 174
   Pom-pom, 160
   Porcelain, 104
   Teddy-bear, 160
   Zebra, 170
Crinoid, 74, 76, 160, 176
Cryptobenthic fish, 210
Cuttlefish, 206
   Broadclub 58
   Flamboyant, 150

Darwin, Charles, 10, 210
Darwin's Paradox, 10, 210
Destructive fishing, blast, cyanide, reef gleaning, 204
Dimorphism, sexual, 120, 130
Dolphin, 54
Dottyback, 38

Eel, Garden, 72
   Moray, 78, 102, 124, 181, 206
   Sand, 48
El Niño, 8
Ecosystem, 126, 164, 174, 178, 194, 196, 202, 208, 210

Filefish, 144
Fish, schools and shoals, 18, 22, 24, 26, 30, 52, 78, 86, 92, 96, 100, 110, 126, 136, 182
Fish, curiosity of, 30, 52, 86
Flatworm, 18, 140, 146
Frogfish, 34, 38, 82, 96, 98, 184
Fusilier, 30, 86, 96, 136, 182
Fisheye lens, 24, 218

Gastropod, Cone shell, 62
   Cowrie, 80
   Opisthobranch, 80
   Ovulid, 80
   See also nudibranch
Geology, 210
Goatfish, 100
Goby, 98, 104, 122, 164, 210
Gorgonian, 122
Gorgonian wrapper – see Tiger anemone
Great Barrier Reef, 8, 10, 170, 202
Grouper, 22, 34, 86, 204
   Red coral, 30
   Highfin, 38
   Humpback, 172

Homochromy, 50, 68
Hybridisation, 132
Hydroid, 68, 76, 158

Invasive species, 122, 124, 126, 208

Jackfish, 92, 100
Jellyfish, 32, 194, 196
   Golden, 78
   Moon, 78

Kleptoparasitism, 160, 176
Komba, 72, 212
Komodo dragon, 192
Kusu Ridge, 2

Lizardfish, 98
Lobster, 204
   Crinoid squat, 74, 176
   Hairy squat, 68
   Painted spiny, 74
   Slipper, 74
   Squat, 104

Mandarinfish, 130
Manta ray, 30, 52, 86, 92, 104, 138, 178, 196
Mantle – see Gastropod
Mimicry, 18, 34, 38, 46, 50, 80, 100, 140, 146, 148, 184
   aggressive, 34
   Batesian, 18, 34,
   dynamic, 148
Mola mola – see Sunfish

Nematocyst, 10
Night dive, 66, 78, 112
Notochord, 82
Nudibranch, 80, 82, 116, 118, 124, 170, 206
   life cycle, reproduction, and protandry, 114, 116

Octopus, 34, 148, 204
    Blue-ring, 60
    Coconut or veined, 152, 154
    Mimic, 148
    Mototi, 60

Parasite, 32, 42, 46, 52, 122, 166, 170, 178, 182
    tongue, 166
Parrotfish, 16, 22, 210
Pearlfish, 110
Photosynthesis, 10, 210
Phytoplankton, 86, 210
Pigment, patterns, 38, 112, 130
Pilot fish, 29, 138
Pipefish, 48
    Ornate ghost, 128, 210
    Slender and embryo cannibalism, 128
    Thread or pipe dragon, 50
Pollution, 18, 50, 196, 202, 206, 208
Pufferfish, 32, 34, 126, 178
Pygmy pipe dragon, see pipefish

Raja Ampat, 18
Rebreather, and silence, 22, 24, 26, 30, 86, 194

Saron shrimp – see Shrimp, Marbled,
Scallop, 174
Scorpionfish, 36, 96, 98, 124
    Lionfish, 38, 122, 148, 208
    Rhinopias, 36
Sea cucumber, 110, 120, 204

Seahorse, 48, 128
    Pygmy, 50
Seamoth, 48, 128
Sea pen, 104, 120
Sea snake, Golden, 148, 190
Sea spider, 68
Sea star, 106, 120, 126, 176
    Blue, Multi-pore, Red, regeneration, 106
    Crinoid (or feather), 74, 76, 176
    Brittle, 110
    fissiparity, and reproduction, 120
Sea urchin, 36, 42, 74, 106, 110, 124, 126, 170, 172, 194
Shark, 42, 52, 78, 92, 112, 170, 182, 190
    Blacktip reef, 30
    Carpet (or wobbegong), 90, 96
    Whale, 54, 86, 104, 186
Shell – see Gastropod
Shrimp, 34, 64, 106, 110, 150, 156, 178, 204
    Algae, 70
    Bumblebee, 62, 64
    Coleman, 170
    Crinoid, 160, 176
    Decorator, 106
    Emperor, 116
    Goby, 164
    Hairy, 70
    Harlequin, 106
    Leopard, 64, 122

Mantis, 96, 108
    Marbled, 66
    Skeleton, 70
    Snapping, 164, 176
    Tiger, 64, 80
Soldierfish, 110
Sponge, 14, 30, 46, 68, 82, 86, 96, 110, 120, 144, 158, 194, 206
Squid, 34, 78, 108,
    Bigfin, 112
    Bobtail, 156
Stargazer, 98
Stingray, 52, 100
Sulawesi 2, 8, 124, 148, 202
    '50 Reefs Initiative' 8
Sunfish, evolution and biology, 32
Sweepers, 22, 24, 78
Sweetlips, 24, 78, 146
Symbiont, 160, 170, 174, 176
Symbiosis, 10, 38, 156, 160, 164, 166, 172, 174, 176, 178, 202, 210
    commensalism, mutualism, parasitism, 164

Tectonic, 2, 170, 212
Tentacle, 10, 14, 68, 78, 80, 124, 150, 156, 166
Tools, use of, 152, 160
Toxin, 44, 60, 62, 78, 82, 148, 150, 192
Trevally, 136, 182
    Big-eye, 92
    Dart, 30
    Giant, 108

Triggerfish, 102, 144
    Black durgon, 42
    Red tooth, 42
    Titan, 42
Trunkfish – see Boxfish
Turtle, 170
    Green, 52, 194
    Hawksbill, 196

Venom, 36, 60, 62, 70, 96, 146, 148, 170, 190, 192, 208
Volcano, 2, 22, 190, 210, 212
Volcanic deposits, 148
    sand, 72, 124, 148
    rock, 210

Wallace, Alfred Russell, 10, 206
West Papua, 2, 132, 202
Worm, 34, 74, 80, 196
    Bobbit, 72, 96
    Christmas tree, 174
    Plume, 46, 144, 184
Wrasse, 32, 102, 178, 204, 210
    Cleaner, 18, 38, 46, 182
    Flasher, 130, 132
    Napoleon, 100

Zoanthid, 158

T - #0053 - 100924 - C256 - 216/280/15 [17] - CB - 9780367428167 - Gloss Lamination